Mathematical Methods in Applied Sciences

Mathematical Methods in Applied Sciences

Special Issue Editor
Luigi Rodino

MDPI • Basel • Beijing • Wuhan • Barcelona • Belgrade

Special Issue Editor
Luigi Rodino
University of Torino
Italy

Editorial Office
MDPI
St. Alban-Anlage 66
4052 Basel, Switzerland

This is a reprint of articles from the Special Issue published online in the open access journal *Mathematics* (ISSN 2227-7390) from 2019 to 2020 (available at: https://www.mdpi.com/journal/mathematics/special_issues/mmas).

For citation purposes, cite each article independently as indicated on the article page online and as indicated below:

LastName, A.A.; LastName, B.B.; LastName, C.C. Article Title. *Journal Name* **Year**, *Article Number*, Page Range.

ISBN 978-3-03928-496-2 (Pbk)
ISBN 978-3-03928-497-9 (PDF)

© 2020 by the authors. Articles in this book are Open Access and distributed under the Creative Commons Attribution (CC BY) license, which allows users to download, copy and build upon published articles, as long as the author and publisher are properly credited, which ensures maximum dissemination and a wider impact of our publications.

The book as a whole is distributed by MDPI under the terms and conditions of the Creative Commons license CC BY-NC-ND.

Contents

About the Special Issue Editor vii

Preface to "Mathematical Methods in Applied Sciences" ix

Ioannis Dassios, Andrew Keane and Paul Cuffe
Calculating Nodal Voltages Using the Admittance Matrix Spectrum of an Electrical Network
Reprinted from: *Mathematics* **2019**, *7*, 106, doi:10.3390/math7010106 1

Şeyda Gür, Tamer Eren and Hacı Mehmet Alakaş
Surgical Operation Scheduling with Goal Programming and Constraint Programming: A Case Study
Reprinted from: *Mathematics* **2019**, *7*, 251, doi:10.3390/math7030251 7

Chunlei Ruan
Chebyshev Spectral Collocation Method for Population Balance Equation in Crystallization
Reprinted from: *Mathematics* **2019**, *7*, 317, doi:10.3390/math7040317 31

Özlem Kaçmaz, Haci Mehmet Alakaş and Tamer Eren
Shift Scheduling with the Goal Programming Method: A Case Study in the Glass Industry
Reprinted from: *Mathematics* **2019**, *7*, 561, doi:10.3390/math7060561 43

Aleksandras Krylovas, Natalja Kosareva, Rūta Dadelienė and Stanislav Dadelo
Evaluation of Elite Athletes Training Management Efficiency Based on Multiple Criteria Measure of Conditioning Using Fewer Data
Reprinted from: *Mathematics* **2020**, *8*, 66, doi:10.3390/math8010066 65

Awatif Jahman Alqarni, Azmin Sham Rambely and Ishak Hashim
Dynamic Modelling of Interactions between Microglia and Endogenous Neural Stem Cells in the Brain during a Stroke
Reprinted from: *Mathematics* **2020**, *8*, 132, doi:10.3390/math8010132 82

Fatin Amani Mohd Ali, Samsul Ariffin Abdul Karim, Azizan Saaban, Mohammad Khatim Hasan, Abdul Ghaffar, Kottakkaran Sooppy Nisar and Dumitru Baleanu
Construction of Cubic Timmer Triangular Patches and its Application in Scattered Data Interpolation
Reprinted from: *Mathematics* **2020**, *8*, 159, doi:10.3390/math8020159 103

About the Special Issue Editor

Luigi Rodino, Professor, University of Torino, Italy. Education: degree in Mathematics, University of Torino 1971; post-doc 1972–75: University of Lund (Sweden), Institut Mittag Leffler (Sweden), University of Princeton (USA). Professor at the University of Torino starting from 1976, Director Department of Mathematics 1988–91, President Faculty in Mathematics for Finance and Insurance 2006–2009. Coordinator International Research Projects NATO and UNESCO 1995–2005. Editor-in-Chief of two international journals, member of the Editorial Committee of 22 journals. President ISAAC, International Society Analysis Applications Computation, 2013–2016. Main research fields: Partial Differential Equations and Fourier Analysis. Author of 144 papers, 5 monographs, 15 edited volumes, 800 reviews; 17 Ph.D. students.

Preface to "Mathematical Methods in Applied Sciences"

"The book of Nature is written in the language of Mathematics". This famous statement of Galileo Galilei (1564–1642) may serve as introduction to this Special Issue. Of course, over the course of four centuries, Mathematics grew enormously, not only in the direction of differential calculus, but thanks to new disciplines, as Probability, Statistics, and Computer-Assisted Numerical Analysis. Simultaneously, the range of applications extended from Mathematical Physics to other fields, such as Biology and Chemistry, Medicine and Public Health, Economy and Industry, and the Social Sciences. The present Special Issue of Mathematics consists of seven articles on mathematical models, expressed in terms of different mathematical disciplines, and addressed to Applied Sciences. New mathematical results are present as well, but emphasis is placed on the effectiveness of mathematical models on different aspects of modern life. We address readers to the seven articles for a detailed presentation of the different topics, and we limit ourselves here to giving an overview of some of the relevant achievements in the present volume. Concerning first Medicine and Public Health, in connection with Social Sciences: the study of the brain cells during a stroke is studied, with particular attention to the interactions between microglia and neural stem cells; training management efficiency is considered for elite athletes, aiming to achieve their peak performance during the main competitions; an optimal surgical operation scheduling is discussed, considering the hospital's sensitive and expensive equipment. Concerning Industry and Economy: nodal voltages are calculated in a meshed network, as fundamental to electric engineering; a case study on the glass industry is presented to emphasize the relevance of resource utilization and management for businesses. Other relevant contributions concern population balance equation in crystallization and problems in Numerical Analysis, in particular scattered data interpolation, spectral collocation methods, and the use of the eigenvalues and eigenvectors of the Laplacian matrix. In the whole, the volume is an excellent witness of the relevance of Mathematical Methods in Applied Sciences.

Luigi Rodino
Special Issue Editor

Article

Calculating Nodal Voltages Using the Admittance Matrix Spectrum of an Electrical Network

Ioannis Dassios *, Andrew Keane and Paul Cuffe

School of Electrical and Electronic Engineering, University College Dublin, Dublin 4, Ireland; Andrew.Keane@ucd.ie (A.K.); paul.cuffe@ucd.ie (P.C.)
* Correspondence: ioannis.dassios@ucd.ie

Received: 29 November 2018; Accepted: 18 January 2019; Published: 20 January 2019

Abstract: Calculating nodal voltages and branch current flows in a meshed network is fundamental to electrical engineering. This work demonstrates how such calculations can be performed using the eigenvalues and eigenvectors of the Laplacian matrix which describes the connectivity of the electrical network. These insights should permit the functioning of electrical networks to be understood in the context of spectral analysis.

Keywords: Laplacian matrix; power flow; admittance matrix; voltage profile

1. Introduction

Electrical power system calculations rely heavily on the bus admittance matrix, Y_{bus}, which is a Laplacian matrix weighted by the complex-valued admittance of each branch in the network. It is well established that the eigenvalues and eigenvectors (deemed the *spectrum*) of a Laplacian matrix encode meaningful information about a network's structure [1]. Recent work in [2,3] indicates that, in electrical networks, this spectrum can be directly related to nodal voltages and branch current flows. The purpose of the present paper is to *clarify* the derivations provided in [3]. The scope of the present work is narrowly theoretical: Linear algebra is used to articulate the correct relationship between the variables treated in [3].

Notwithstanding these modest ambitions, a key motivation for the present work is to begin to link power flow analysis with the mature literature [4] on spectral graph theory. Extant efforts to apply spectral graph theory to electrical networks are scarce, but include [5,6]. The use of graph theory more generally in this role is reviewed in [7,8]. Notably, simplistic topological approaches do not properly account for the physical realities of electrical power flow, and can thereby fail to identify the critical components in an electrical network [9–11]. The present work seeks to articulate one particular linkage between spectral graph theory and circuit theory, which may offer new ways to understand how power flows in meshed electrical networks.

The rest of this paper is organized as follows: In Section 2 we establish the necessary preliminaries, including electrical flow basics and notation. The main results are presented in Section 3.

2. Preliminaries

2.1. Electrical Flow Basics and Notation

Ohm's law linearly relates the current flowing through an edge in a circuit with the voltage difference between the nodes that the edge connects. Specifically, $I_{kj} = \frac{\Delta V_{kj}}{Z_{kj}}$, and $\sum_{j=1}^{N} I_{kj} = F_k$, $k, j = 1, 2, \ldots, N$, where I_{kj} is the current passing from the k-th node to the j-th in a (typically sparsely connected) network of N nodes, $\Delta V_{kj} = V_k - V_j$ is the voltage difference between the k-th node and the j-th, $Z_{kj} = Z_{jk}$ is branch impedance and F_k are complex-valued net current injections or withdrawals.

From the above notation we arrive easily at $\sum_{j=1}^{N} \frac{\Delta V_{kj}}{Z_{kj}} = F_k$, $\forall k = 1, 2, \ldots, N$, i.e., $\sum_{j=1}^{N} \frac{(V_k - V_j)}{Z_{kj}} = F_k$, $\forall k = 1, 2, \ldots, N$, or, equivalently,

$$V_k \sum_{j=1}^{N} \frac{1}{Z_{kj}} - \sum_{j=1}^{N} \frac{V_j}{Z_{kj}} = F_k, \quad \forall k = 1, 2, \ldots, N. \quad (1)$$

In the article we will denote with δ_{ij} for the Kronecker delta, i.e., $\delta_{ii} = 1$ and $\delta_{ij} = 0$ for $i \neq j$. With $\bar{u}$ we will denote the complex conjugate of u, and with ^{T}, * the conjugate transpose, and conjugate transpose tensor respectively.

2.2. An Exemplary Electrical Network

To provide some context, a nation-spanning electrical power system is shown in Figure 1. This diagram of the `nesta_case2224_edin` test system [12] was created using the techniques described in [13], which uses electrical distances measures, rather than physical geography, to positions nodes. Note the relative spareseness of its connective struture, and how lower nominal voltage levels (< 143 kV) correspond to more tree-like structures. This network of 2224 nodes supplies a total load of up to 60 GW, supplied from 378 different generating sites.

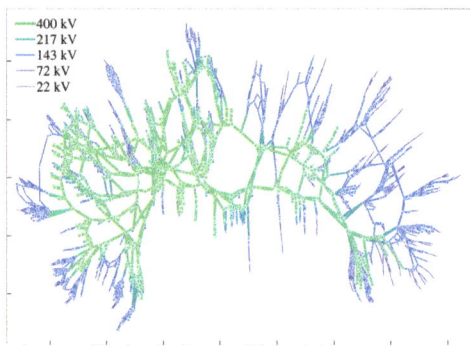

Figure 1. This diagram shows the `nesta_case2224_edin` test power system

3. Derivations

In this section, first we rewrite (1) in matrix form and define the relevant Laplacian matrix. Then we provide a formula which explicitly relates the voltage differences to the eigenvalues and eigenvectors of the Laplacian matrix for meshed electrical networks. We can state now the following theorem.

Theorem 1. *Consider an electrical network with branch currents I_{kj}, $\forall\, k, j = 1, 2, \ldots, N$ passing from node k to node j, a complex impedance describing each branch $Z_{kj} = Z_{jk}$, and F_k being the complex-valued net current flow at each bus with $\sum_{k=1}^{N} F_k = 0$. Then the voltage difference ΔV_{mn} between two arbitrary nodes m and n is given by:*

$$\Delta V_{mn} = \sum_{j=2}^{N} \left[\frac{u_{mj} - u_{nj}}{\lambda_j} \left(\sum_{k=1}^{N} \bar{u}_{kj} F_k \right) \right] \quad (2)$$

where λ_k, $k = 2, 3, \ldots, N$ are the non-zero eigenvalues of the G matrix (equivalent to the Y_{bus} matrix in the power systems context) which describes the connectivity of the electrical network:

$$G = [\delta_{kr} (\sum_{j=1}^{N} \frac{1}{Z_{kj}}) + (\delta_{kr} - 1) \frac{1}{Z_{kr}}]_{k=1,2,\ldots,N}^{r=1,2,\ldots,N} \quad (3)$$

and $\begin{bmatrix} u_{1k} & u_{2k} & \cdots & u_{Nk} \end{bmatrix}^T$ is an eigenvector of the eigenvalue λ_k.

Proof. For $k = 1, 2, \ldots, N$, Equation (1) can be written as

$$\begin{bmatrix} V_1 \sum_{j=1}^N \frac{1}{Z_{1j}} - \sum_{j=1}^N \frac{V_j}{Z_{1j}} \\ V_2 \sum_{j=1}^N \frac{1}{Z_{2j}} - \sum_{j=1}^N \frac{V_j}{Z_{2j}} \\ \vdots \\ V_N \sum_{j=1}^N \frac{1}{Z_{Nj}} - \sum_{j=1}^N \frac{V_j}{Z_{Nj}} \end{bmatrix} = \begin{bmatrix} F_1 \\ F_2 \\ \vdots \\ F_N \end{bmatrix},$$

or, equivalently,

$$\begin{bmatrix} V_1 \sum_{j=1}^N \frac{1}{Z_{1j}} - \frac{V_1}{Z_{11}} - \cdots - \frac{V_N}{Z_{1N}} \\ V_2 \sum_{j=1}^N \frac{1}{Z_{2j}} - \frac{V_1}{Z_{21}} - \cdots - \frac{V_N}{Z_{2N}} \\ \vdots \\ V_N \sum_{j=1}^N \frac{1}{Z_{Nj}} - \frac{V_1}{Z_{N1}} - \cdots - \frac{V_N}{Z_{NN}} \end{bmatrix} = \begin{bmatrix} F_1 \\ F_2 \\ \vdots \\ F_N \end{bmatrix},$$

or, equivalently, by setting $V = \begin{bmatrix} V_i \end{bmatrix}_{i=1,2,\ldots,N}$, $F = \begin{bmatrix} F_i \end{bmatrix}_{i=1,2,\ldots,N}$, and

$$G = \begin{bmatrix} \sum_{j=1}^N \frac{1}{Z_{1j}} - \frac{1}{Z_{11}} & \cdots & -\frac{1}{Z_{1N}} \\ -\frac{1}{Z_{21}} & \cdots & -\frac{1}{Z_{2N}} \\ \vdots & \ddots & \vdots \\ -\frac{1}{Z_{N1}} & \cdots & \sum_{j=1}^N \frac{1}{Z_{Nj}} - \frac{1}{Z_{NN}} \end{bmatrix},$$

We arrive at $GV = F$. We observe that if G_{kj}, $k, j = 1, 2, \ldots, N$ is an element of G, then for $k = j$, $G_{kk} = \sum_{j=1}^N \frac{1}{Z_{kj}} - \frac{1}{Z_{kk}}$ and for $k \neq j$, $G_{kj} = -\frac{1}{Z_{kj}}$. Hence G is given by (3). We will refer to G as the Laplacian matrix. Note that the rows of G sum to zero, i.e., the matrix has the zero eigenvalue (see [3,14]). The algebraic multiplicity of the zero eigenvalue in the Laplacian is the number of connected components in the network. In the power systems case we deal with only one network which means the algebraic multiplicity of the zero eigenvalue is one. Since the matrix G is symmetric it can be written in the following form:

$$G = PDP^*,$$

where $P = [u_{kj}]_{k=1,2,\ldots,N}^{j=1,2,\ldots,N}$, P^* is the conjugate transpose of P such that PP^* is the $N \times N$ identity matrix and D is the diagonal matrix $D = diag\{0, \lambda_2, \lambda_3, \ldots, \lambda_N\}$. By applying the above expression into the system we get:

$$PDP^*V = F,$$

and since P^* is the inverse of P we have:

$$DP^*V = P^*F,$$

or, equivalently,

$$\begin{bmatrix} 0 \\ \lambda_2 \sum_{k=1}^N \bar{u}_{k2} V_k \\ \lambda_3 \sum_{k=1}^N \bar{u}_{k3} V_k \\ \vdots \\ \lambda_N \sum_{k=1}^N \bar{u}_{kN} V_k \end{bmatrix} = \begin{bmatrix} \sum_{k=1}^N \bar{u}_{k1} F_k \\ \sum_{k=1}^N \bar{u}_{k2} F_k \\ \sum_{k=1}^N \bar{u}_{k3} F_k \\ \vdots \\ \sum_{k=1}^N \bar{u}_{kN} F_k \end{bmatrix}.$$

Let $\mathbf{1}_N$ be a column vector that contains exactly N 1's. From the fact that every row of G sums to zero we have the eigenspace of the zero eigenvalue. Indeed $G \cdot \mathbf{1}_N = 0 \cdot \mathbf{1}_N$ which means that $<\mathbf{1}_N>$ is the eigenspace of the zero eigenvalue. Hence there exist $c \in \mathbb{C}$ such that

$$\begin{bmatrix} u_{i1} \end{bmatrix}_{i=1,2,\ldots,N} = c \cdot \mathbf{1}_N \tag{4}$$

From (4) $u_{k1} = c$, $\forall k = 1, 2, \ldots, N$, or, equivalently, $\bar{u}_{k1} = \bar{c}$, $\forall k = 1, 2, \ldots, N$. In addition, $\sum_{k=0}^{N} F_k = 0$. Hence, $\sum_{k=1}^{N} \bar{u}_{k1} F_k = \bar{c} \sum_{k=1}^{N} F_k = 0$. By ignoring the first row of each column of the above expression we get:

$$\begin{bmatrix} \lambda_2 \sum_{k=1}^{N} \bar{u}_{k2} V_k \\ \lambda_3 \sum_{k=1}^{N} \bar{u}_{k3} V_k \\ \vdots \\ \lambda_N \sum_{k=1}^{N} \bar{u}_{kN} V_k \end{bmatrix} = \begin{bmatrix} \sum_{k=1}^{N} \bar{u}_{k2} F_k \\ \sum_{k=1}^{N} \bar{u}_{k3} F_k \\ \vdots \\ \sum_{k=1}^{N} \bar{u}_{kN} F_k \end{bmatrix}.$$

Which can be rewritten in the following form:

$$\begin{bmatrix} \sum_{k=1}^{N} \bar{c} V_k \\ \lambda_2 \sum_{k=1}^{N} \bar{u}_{k2} V_k \\ \lambda_3 \sum_{k=1}^{N} \bar{u}_{k3} V_k \\ \vdots \\ \lambda_N \sum_{k=1}^{N} \bar{u}_{kN} V_k \end{bmatrix} = \begin{bmatrix} \sum_{k=1}^{N} \bar{c} V_k \\ \sum_{k=1}^{N} \bar{u}_{k2} F_k \\ \sum_{k=1}^{N} \bar{u}_{k3} F_k \\ \vdots \\ \sum_{k=1}^{N} \bar{u}_{kN} F_k \end{bmatrix}.$$

If we set $\Lambda = \mathrm{diag}\{\lambda_i\}_{2 \le i \le N}$, $U = \begin{bmatrix} \bar{u}_{ij} \end{bmatrix}_{i=1,2,\ldots,N}^{j=1,2,\ldots,N}$, we have

$$\begin{bmatrix} 1 & 0_{N-1}^T \\ 0_{N-1} & \Lambda \end{bmatrix} \begin{bmatrix} \bar{c} \cdot \mathbf{1}_N^T \\ \bar{U} \end{bmatrix} V = \begin{bmatrix} \sum_{k=1}^{N} \bar{c} V_k \\ \sum_{k=1}^{N} \bar{u}_{k2} F_k \\ \sum_{k=1}^{N} \bar{u}_{k3} F_k \\ \vdots \\ \sum_{k=1}^{N} \bar{u}_{kN} F_k \end{bmatrix},$$

or, equivalently,

$$\begin{bmatrix} \bar{c} \cdot \mathbf{1}_N^T \\ \bar{U} \end{bmatrix} V = \begin{bmatrix} 1 & 0_{N-1}^T \\ 0_{N-1} & \Lambda^{-1} \end{bmatrix} \begin{bmatrix} \sum_{k=1}^{N} \bar{c} V_k \\ \sum_{k=1}^{N} \bar{u}_{k2} F_k \\ \sum_{k=1}^{N} \bar{u}_{k3} F_k \\ \vdots \\ \sum_{k=1}^{N} \bar{u}_{kN} F_k \end{bmatrix},$$

or, equivalently,

$$V = \begin{bmatrix} c \cdot \mathbf{1}_N & U \end{bmatrix} \begin{bmatrix} \sum_{k=1}^{N} \bar{c} V_k \\ \frac{1}{\lambda_2} \sum_{k=1}^{N} \bar{u}_{k2} F_k \\ \frac{1}{\lambda_3} \sum_{k=1}^{N} \bar{u}_{k3} F_k \\ \vdots \\ \frac{1}{\lambda_N} \sum_{k=1}^{N} \bar{u}_{kN} F_k \end{bmatrix}$$

or, equivalently,

$$V = \begin{bmatrix} c\bar{c} \sum_{k=1}^{N} V_k + \sum_{j=2}^{N} \frac{u_{1j}}{\lambda_j} \sum_{k=1}^{N} \bar{u}_{kj} F_k \\ c\bar{c} \sum_{k=1}^{N} V_k + \sum_{j=2}^{N} \frac{u_{2j}}{\lambda_j} \sum_{k=1}^{N} \bar{u}_{kj} F_k \\ \vdots \\ c\bar{c} \sum_{k=1}^{N} V_k + \sum_{j=2}^{N} \frac{u_{Nj}}{\lambda_j} \sum_{k=1}^{N} \bar{u}_{kj} F_k \end{bmatrix}$$

or, equivalently, for $b_j = \sum_{k=1}^{N} \bar{u}_{kj} F_k$:

$$V = \begin{bmatrix} \sum_{k=1}^{N} \bar{u}_{k1} V_k + \sum_{j=2}^{N} \frac{u_{1j}}{\lambda_j} b_j \\ \sum_{k=1}^{N} \bar{u}_{k1} V_k + \sum_{j=2}^{N} \frac{u_{2j}}{\lambda_j} b_j \\ \vdots \\ \sum_{k=1}^{N} \bar{u}_{k1} V_k + \sum_{j=2}^{N} \frac{u_{Nj}}{\lambda_j} b_j \end{bmatrix}.$$

Let V_m, V_n be two arbitrary nodal voltages, i.e.,

$$V_m = c\bar{c} \sum_{k=1}^{N} V_k + \sum_{j=2}^{N} \frac{u_{mj}}{\lambda_j} b_j,$$
$$V_n = c\bar{c} \sum_{k=1}^{N} V_k + \sum_{j=2}^{N} \frac{u_{nj}}{\lambda_j} b_j.$$

Then, the difference between them is given by

$$\Delta V_{mn} = c\bar{c} \sum_{k=1}^{N} V_k + \sum_{j=2}^{N} \frac{u_{mj}}{\lambda_j} b_j - c\bar{c} \sum_{k=1}^{N} V_k - \sum_{j=2}^{N} \frac{u_{nj}}{\lambda_j} b_j,$$

or, equivalently,

$$\Delta V_{mn} = \sum_{j=2}^{N} \frac{u_{mj} - u_{nj}}{\lambda_j} b_j,$$

or, equivalently,

$$\Delta V_{mn} = \sum_{j=2}^{N} \left[\frac{u_{mj} - u_{nj}}{\lambda_j} \left(\sum_{k=1}^{N} \bar{u}_{kj} F_k \right) \right].$$

□

4. Conclusions

This work has clarified the relationship between the admittance matrix spectrum, the current inflows & withdrawals prevailing in an electrical network and the resulting nodal voltage profile. Applying these spectral relationships to practical electrical engineering problems is left to future work.

Author Contributions: Methodology, I.D.; Formal Analysis, I.D. and P.C. and A.K.; Writing—Original Draft Preparation, I.D.; Writing—Review and Editing, I.D. and P.C.; Visualization, P.C.; Supervision, P.C. and A.K.

Funding: This material is supported by the Science Foundation Ireland (SFI), by funding Ioannis Dassios under Investigator Programme Grant No. SFI/15/IA/3074; and A. Keane and P. Cuffe under the SFI Strategic Partnership Programme Grant Number SFI/15/SPP/E3125. The opinions, findings and conclusions or recommendations expressed in this material are those of the authors and do not necessarily reflect the views of the Science Foundation Ireland.

Conflicts of Interest: The authors declare no conflict of interest.

References

1. Mohar, B.; Alavi, Y.; Chartrand, G.; Oellermann, O. The Laplacian spectrum of graphs. *Graph Theory Comb. Appl.* **1991**, *2*, 12.
2. Rubido, N.; Grebogi, C.; Baptista, M.S. Structure and function in flow networks. *EPL (Europhys. Lett.)* **2013**, *101*, 68001. [CrossRef]
3. Rubido, N.; Grebogi, C.; Baptista, M.S. General analytical solutions for DC/AC circuit-network analysis. *Eur. Phys. J. Spec. Top.* **2017**, *226*, 1829–1844. [CrossRef]
4. Chung, F.R.; Graham, F.C. *Spectral Graph Theory*; Number 92; American Mathematical Society: Providence, RI, USA, 1997.
5. Edström, F. On eigenvalues to the Y-bus matrix. *Int. J. Electr. Power Energy Syst.* **2014**, *56*, 147–150. [CrossRef]

6. Edström, F.; Söder, L. On spectral graph theory in power system restoration. In Proceedings of the 2011 2nd IEEE PES International Conference and Exhibition on Innovative Smart Grid Technologies (ISGT Europe), Manchester, UK, 5–7 December 2011; pp. 1–8.
7. Pagani, G.A.; Aiello, M. The power grid as a complex network: A survey. *Phys. A Stat. Mech. Its Appl.* **2013**, *392*, 2688–2700. [CrossRef]
8. Sun, K. Complex networks theory: A new method of research in power grid. In Proceedings of the 2005 IEEE/PES Transmission and Distribution Conference and Exhibition: Asia and Pacific, Dalian, China, 18 August 2005; pp. 1–6.
9. Hines, P.; Cotilla-Sanchez, E.; Blumsack, S. Do topological models provide good information about electricity infrastructure vulnerability? *Chaos Interdiscip. J. Nonlinear Sci.* **2010**, *20*, 033122. [CrossRef] [PubMed]
10. Verma, T.; Ellens, W.; Kooij, R.E. Context-independent centrality measures underestimate the vulnerability of power grids. *arXiv* 2013, arXiv:1304.5402.
11. Cuffe, P. A comparison of malicious interdiction strategies against electrical networks. *IEEE J. Emerg. Sel. Top. Circuits Syst.* **2017**, *7*, 205–217. [CrossRef]
12. Coffrin, D.; Gordon, D.; Scott, P. Nesta the nicta energy system test case archive. *arXiv* 2014, arXiv:1411.0359.
13. Cuffe, P.; Keane, A. Visualizing the electrical structure of power systems. *IEEE Syst. J.* **2017**, *11*, 1810–1821. [CrossRef]
14. Dassios, I.K.; Cuffe, P.; Keane, A. Visualizing voltage relationships using the unity row summation and real valued properties of the F_{LG} matrix. *Electr. Power Syst. Res.* **2016**, *140*, 611–618. [CrossRef]

© 2019 by the authors. Licensee MDPI, Basel, Switzerland. This article is an open access article distributed under the terms and conditions of the Creative Commons Attribution (CC BY) license (http://creativecommons.org/licenses/by/4.0/).

Article

Surgical Operation Scheduling with Goal Programming and Constraint Programming: A Case Study

Şeyda Gür, Tamer Eren * and Hacı Mehmet Alakaş

Department of Industrial Engineering, Faculty of Engineering, Kirikkale University, 71450 Kirikkale, Turkey; seydaaa.gur@gmail.com (Ş.G.); hmalagas@gmail.com (H.M.A.)
* Correspondence: tamereren@gmail.com; Tel.: +90-3183574242

Received: 13 November 2018; Accepted: 19 February 2019; Published: 11 March 2019

Abstract: The achievement of health organizations' goals is critically important for profitability. For this purpose, their resources, materials, and equipment should be efficiently used in the services they provide. A hospital has sensitive and expensive equipment, and the use of its equipment and resources needs to be balanced. The utilization of these resources should be considered in its operating rooms, as it shares both expense expenditure and revenue generation. This study's primary aim is the effective and balanced use of equipment and resources in hospital operating rooms. In this context, datasets from a state hospital were used via the goal programming and constraint programming methods. According to the wishes of hospital managers, three scenarios were separately modeled in both methods. According to the obtained results, schedules were compared and analyzed according to the current situation. The hospital-planning approach was positively affected, and goals such as minimization cost, staff and patient satisfaction, prevention over time, and less use were achieved.

Keywords: scheduling; operating room scheduling; goal programming; constraint programming; state hospital

1. Introduction

Factors such as resource efficiency, the hospital's financial situation, and the needs of the staff play an important role in hospitals, which belong to the service industry and have complex processes. Operating rooms play a large role in hospital budgets, as they contain sensitive and expensive equipment, and they have been shown to constitute up to 40% of hospital expenses and account for two-thirds of hospital income [1]. In these units, where surgeons perform various lifesaving medical interventions, human life is critical.

Scheduling activities determine the order and time in which jobs are performed. The capacity of resources must be taken into account when determining the start and completion times of these activities. The effective use of limited resources in operating rooms comes to the fore. When scheduling operating room activities, hospitals should allocate resources [2].

There are many studies in the literature related to operating room scheduling. Ozkarahan [3] looked at increasing the quality of service provided in healthcare institutions with increasing demands and developed a scheduling model that met the needs of hospitals to a great extent. Arenas et al. [4] studied a hospital in Spain, aiming at reducing patient waiting time, while at the same time balancing resources in a hospital. Blake and Donald's [5] study emphasized the seriousness of operating room cost in hospitals, while Blake and Carter [6] addressed the issue of resource allocation in hospitals, using the goal programming method in their study to keep service costs constant so that they could set goals for the hospital to balance their expenses with the care that they provide.

Kharraja et al. [7] pointed out that, in the schedules created for operating rooms in hospitals, strategies should be directed toward increasing profitability. Wullink et al. [8] studied the rate at which the operating rooms could respond to emergency situations by assessing the situations in which emergency cases occurred. Paoletti and Marty [9] evaluated stochastic situations with the Monte Carlo simulation model. Lamiriet et al. [10] considered uncertain situations for operating room planning. Beliën et al. [11] pointed to the inefficiencies of resource use in the operating room with a comprehensive case study. Zhang et al. [12] emphasized minimizing the costs associated with the length of stay of hospital patients.

A few of the main purposes of scheduling activities are to deliver a job on time to the customer, both keeping overtime in the operating room to a minimum and reducing idle times in order to ensure efficient use. When these goals are specifically intended for hospitals, a scheduled procedure in the operating room must be performed on the specified day and at the specified time. Thus, the satisfaction of the patients, referred to as customers, can be met at the highest level. At the same time, the effectiveness of resources is ensured by avoiding idleness and overtime in operating rooms and resources [13].

Fei et al. [14] wanted to balance the use of both hospital surgery rooms and recovery rooms. At the same time, they aimed to minimize overtime hours and balance costs. They analyzed and compared the monthly schedule of a hospital with its actual schedule. Tànfani and Testi [15] aimed to offer a solution to the desired goals of administration and surgeons during the process of scheduling appointments for an operating room. Banditori et al. [16] sought solutions to problems arising from high waiting times for patients on the waiting lists and the inability to homogenous distribution of hospital resources. Cappanera et al. [17] aimed to better balance the workload of surgeons in operating room schedules, distribute resources more efficiently and fairly, and plan more systematically. Eren et al. [18] designed an application to address the problem associated with operating room scheduling. They developed a mathematical model for this application and realized their aims. Xiang [19] modeled the objectives of operating room schedules using ant colony optimization and Pareto sets. Abedini et al. [20] in their work addressed the problem of balancing resources between operating room units.

When the above literature was evaluated in detail, it was found that it emphasized that operating room usage should be the most productive, both in terms of time and resources. In this study, a model is proposed for the problems associated with operating room scheduling using integrated goal programming and constraint programming methods and datasets from a state hospital. In schedules created with the integrated goal programming and constraint programming methods, it was attempted to minimize deviations for these purposes, and a flexible model was established.

Looking at studies from different application fields in the literature (e.g., References [21–27]), the goal programming method allows decision-makers to simultaneously realize different goals [28,29]. The most important features that distinguish constraint programming from mathematical programming include mathematical constraints or constraints that can be of a logical or symbolic type [30–32]. The integrated goal programming and constraint programming methods, which are used in the solution process, are effective solution tools for researchers. Each objective set in the integrated goal programming and constraint programming methods is defined as a constraint and minimizes deviations from these objectives. A solution can always be obtained using these methods. However, the resulting decision-makers must be satisfied [33]. The constraint programming method can include mathematical constraints, as well as constraints that can be logical or symbolic. Constraint programming problems have structures that contain the definition set, constraints, and purpose of the decision variables, are identified with an appropriate language, and are solved via polynomial. Mathematical modelling is used for the solution of the problem [34]. However, due to the nature of the problem, the integrated goal programming and constraint programming models were developed, as it was desired to simultaneously perform more than one purpose. These methods were integrated with each other in the study.

In the operating room scheduling problem, the aim is to assign operations to the operating rooms in the appropriate time intervals. For this, block times have been defined within one working day, and restrictions have been made for assigning the same expertise to each block. Only the assignment of operations is taken into account, while the ranking of operations is not taken into consideration. This situation is also mentioned in the assumptions. The problem is solved by using the integrated goal programming and constraint programming models.

The study consists of three parts. The first part is the introduction, where scheduling activities are discussed in general, information about operating room scheduling is given, and the importance of this scheduling style is mentioned. Information about the method used in the solution process is also given. In the second part, we discuss a case study. In the third part, the implementation results are examined. The results are also interpreted in general, and suggestions are made for future studies.

2. Method

This study considers some situations that can be encountered in real life, using data from a state hospital. Increasing quality of service with the planned schedule in operating rooms, which are shown to be among the most important units of health institutions, is the main aim of this study. The flowchart of the study is shown in Figure 1 and briefly summarizes the study.

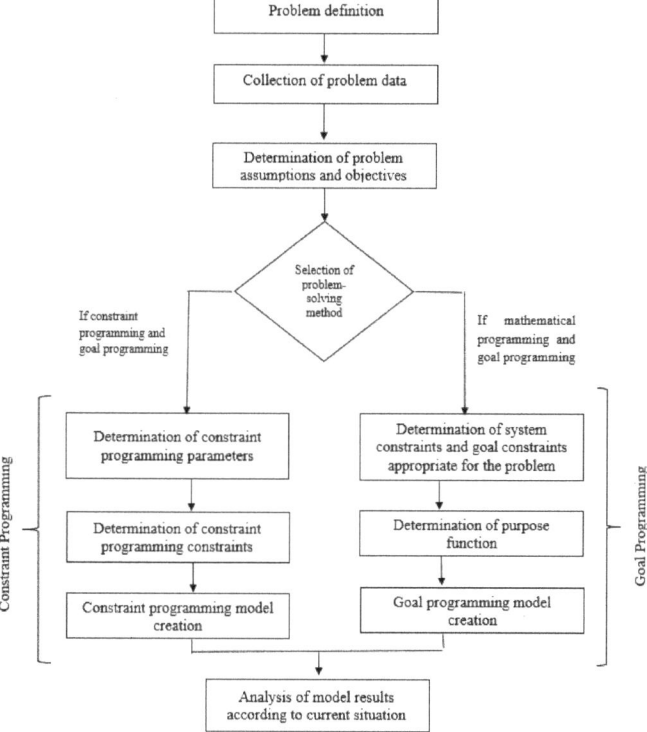

Figure 1. Case study flowchart.

There are various study requirements, such as the presence of equipment and devices that must be available in each operating room. Under various constraints, the assignments of these operations must be done in a systematic manner. Because there are many external factors, the operating room contains much uncertainty in its structure. For this reason, some assumptions are made in operating room scheduling. These assumptions are as follows:

- The number of operations to be performed at the hospital is certain and emergency situations are not considered.
- There are enough staff members and all the needed resources to perform the operations.
- The average duration of each operation was calculated from past data from similar operations previously performed at the hospital. The duration of these operations includes preparation and cleaning times.
- According to the decision of the hospital management committee, working hours at the hospital are 08:00–17:00, and there is one hour between 12:00 and 13:00 for lunch included in these working hours.
- Surgeries are not held on weekends. For this reason, operating rooms are kept closed during this time.
- Only the assignment status of the operations is taken into account on a working day.

With the operating room schedules established under these assumptions, the following are aimed:

- Minimizing the cost of operating room units in the hospital;
- balanced distribution of resources to prevent overtime and less use;
- block scheduling;
- ensuring patient and staff satisfaction is at the highest level; and
- increasing hospital efficiency by getting the highest performance levels from the working staff.

The schedules formed in operating rooms, which are required to be professionally planned in line with these aims, highlight the effectiveness of the units. The data used in the study were taken from a state hospital: the number of operations, the number and capacity of existing operating rooms, the number of experts, and the operating times including preparation and cleaning times, which were prepared by using historical data. Table 1 shows the operations of specialties and the operating times of these operations.

Table 1. Operating times (Op.: operation).

Orthopedics		General Surgery 1		Cardiovascular Surgeon		Plastic Surgery		General Surgery 2		Urology	
No.	Op. Time (min)	No.	Op. Time (min)	No.	Op. Time (min)	No.	Op. Time (min)	No.	Op. Time (min)	No.	Op. Time (min)
Op.1	135	Op.24	25	Op.50	180	Op.66	70	Op.86	50	Op.110	145
Op.2	65	Op.25	65	Op.51	200	Op.67	85	Op.87	54	Op.111	85
Op.3	55	Op.26	120	Op.52	75	Op.68	35	Op.88	100	Op.112	80
Op.4	150	Op.27	200	Op.53	240	Op.69	65	Op.89	85	Op.113	105
Op.5	100	Op.28	95	Op.54	145	Op.70	110	Op.90	100	Op.114	95
Op.6	70	Op.29	35	Op.55	165	Op.71	90	Op.91	120	Op.115	120
Op.7	175	Op.30	140	Op.56	55	Op.72	135	Op.92	90	Op.116	100
Op.8	130	Op.31	115	Op.57	135	Op.73	29	Op.93	95	Op.117	60
Op.9	75	Op.32	60	Op.58	150	Op.74	64	Op.94	60	Op.118	110
Op.10	120	Op.33	45	Op.59	75	Op.75	90	Op.95	105	Op.119	120
Op.11	130	Op.34	120	Op.60	60	Op.76	95	Op.96	75	Op.120	110
Op.12	145	Op.35	115	Op.61	75	Op.77	75	Op.97	110	-	-
Op.13	115	Op.36	145	Op.62	85	Op.78	60	Op.98	100	-	-
Op.14	65	Op.37	80	Op.63	50	Op.79	95	Op.99	125	-	-
Op.15	85	Op.38	45	Op.64	45	Op.80	90	Op.100	105	-	-
Op.16	105	Op.39	70	Op.65	95	Op.81	80	Op.101	73	-	-
Op.17	84	Op.40	60	-	-	Op.82	115	Op.102	95	-	-
Op.18	100	Op.41	145	-	-	Op.83	100	Op.103	135	-	-
Op.19	115	Op.42	95	-	-	Op.84	77	Op.104	78	-	-
Op.20	125	Op.43	200	-	-	Op.85	105	Op.105	120	-	-
Op.21	80	Op.44	170	-	-	-	-	Op.106	140	-	-
Op.22	120	Op.45	165	-	-	-	-	Op.107	115	-	-
Op.23	85	Op.46	125	-	-	-	-	Op.108	77	-	-
-	-	Op.47	117	-	-	-	-	Op.109	115	-	-
-	-	Op.48	200	-	-	-	-	-	-	-	-
-	-	Op.49	210	-	-	-	-	-	-	-	-

According to this, considering the capacities of the operating rooms, it is desired to minimize the use of overtime and less use. At the same time, a flexible model was created to allow block scheduling. Three different scenarios are considered in the study, considering three different situations. There are 120 operations planned in this hospital, which has six specialties. There are eight operating rooms for assignment of these operations. In the model, there are 10 time zones for assignments. Each time zone represents a four-hour working time. In total, it is thought that two time zones equal with one day.

2.1. Creating Surgical Schedule in Operating Rooms with the Goal Programming Method

Within the framework of the above assumptions and objectives, the model was coded and solved with IBM ILOG CPLEX [35]. The goal programming model is given as follows.

Notation of model-forming expressions:

$X_{ijk} = \begin{cases} 1, & \text{if operation i assigned to j operation room in kth time zone} \\ 0, & \text{otherwise} \end{cases}$ $\forall i, \forall j, \forall k$

$Y_{jks} = \begin{cases} 1, & \text{If it has experts s in operation room j in kth time zone} \\ 0, & \text{otherwise} \end{cases}$ $\forall j, \forall k, \forall s$

$Y_{jks} = Z_{js} = \begin{cases} 1, & \text{If it has experts s in operation room j} \\ 0, & \text{otherwise} \end{cases}$ $\forall j, \forall s$

p_i: time of operation i, including cleaning and preparation times $\forall i$
U_{kj}: Used daily time of operating room j in the kth time zone $\forall j, \forall k$
d_s: the set from s experts that prefers to be assigned to the time zone $\forall s$
M: A large number

Deviation variables:

u_{kj}^-: amount of negative deviation to the total available time from operating room j in kth time zone; $\forall j, \forall k$
u_{kj}^+: amount of positive deviation to the total available time from operating room j in kth time zone; $\forall j, \forall k$
k_j^-: amount of negative deviation from a balanced distribution of time zones in operating room j; $\forall j$
k_j^+: amount of positive deviation from a balanced distribution of time zones in operating room j; $\forall j$
r_s^-: amount of negative deviation from the preferred time zone by experts s; $\forall s$
r_s^+: amount of positive deviation from the preferred time zone by experts s; $\forall s$
p_j^-: amount of negative deviation from assigned experts in operating room j; $\forall j$
p_j^+: amount of positive deviation from assigned experts in operating room j. $\forall j$

Constraints and goals:

$$\sum_{k=1}^{r}\sum_{j=1}^{m} X_{ijk} = 1 \quad \forall i \tag{1}$$

$$\sum_{j=1}^{m}\sum_{s=1}^{S} Y_{jks} \leq 6 \quad \forall k \tag{2}$$

$$\sum_{j=1}^{m} Y_{Jks} \leq 6 \quad \forall k, \forall s \tag{3}$$

$$\sum_{i=1}^{n} X_{ijk} \leq M * Y_{jks} \quad \forall i, \forall k, \forall s \tag{4}$$

$$\sum_{k=1}^{r} Y_{jks} \leq M * Z_{js} \quad \forall j, \forall s \tag{5}$$

$$\sum_{i=1}^{n} p_i X_{ijk} + u_{kj}^- - u_{kj}^+ = U_{kj} \quad \forall j, \forall k \tag{6}$$

$$\sum_{s=1}^{S}\sum_{k=1}^{r} Y_{jks} + k_j^- - k_j^+ = 6 \quad \forall j \tag{7}$$

$$\sum_{k \in d_s} Y_{jks} + r_s^- - r_s^+ = 0 \quad \forall s \tag{8}$$

$$\sum_{s=1}^{S} Z_{js} + p_j^- - p_j^+ = 1 \quad \forall j \tag{9}$$

Constraint 1 states that each operation is assigned to the whole operating room and time zones only once. Constraint 2 ensures that at most six different specialties are able to work in all operating rooms at any time zone. Thus, operations can be carried out with the same equipment and technical personnel in the time zone. Constraints 3 and 4 always ensure that each time zone and each specialty be assigned to all operating rooms six times at most. Thus, in different time zones, different experts can be in balanced distribution in all operating rooms. It is possible to prevent the accumulation of an expert in only one operating room. Constraint 5 makes it possible to assign an expert in all operating rooms throughout the entire time zone. Thus, in the operating room separated by the same expert, operations can be carried out without any waiting or interruption due to equipment and technical personnel changes throughout the whole week.

Constraint 6 aims to minimize operating room overtime and less use of the operating room. It ensures to minimize deviations between reachable time and working time in this goal. The purpose function of this goal is as follows:

$$\min \sum_{k=1}^{r} \sum_{j=1}^{m} (u_{kj}^- + u_{kj}^+)$$

Goal Constraint 7 aims to distribute all operations in all operating rooms in a balanced manner. The right side is set to six because, in the problem with the 10 time zones, it is required to assign an average of six time zones to each operating room. Negative and positive deviations are minimized by using this constraint. The purpose function of this goal is as follows:

$$\min \sum_{j=1}^{m} k_j^- + k_j^+$$

Goal Constraint 8 aims to perform the operations of a specialist surgeon in each operating room at the desired time zones. When the constraint is run, it is equal to zero by nothing the time zones in which we do not want to be assigned. In this goal constraint, deviation in the positive direction is minimized. Thus, attempts are made to prevent assignment to undesired time zones. The purpose function of this goal is as follows:

$$\min \sum_{s=1}^{S} r_s^+$$

Goal Constraint 9 is aimed at allocation each operating room for different specialties during the week. Thus, a specialist is working all the time during the week. In this goal constraint, deviation in the positive direction is minimized. The purpose function of this goal is as follows:

$$\min \sum_{j=1}^{m} p_j^+$$

With this mathematical model, three different scenarios were developed. Block scheduling was considered for 120 operations. Block schedules are enacted according to the logic of working certain specialties into specific time zones. The created scenarios are as follows:

Scenario 1. *Block scheduling was worked into this model. In order to prevent overtime and less use, same-expertise operations are assigned to the same time zones. In this scenario, in the objective function, deviation in the negative and positive directions related to using operating room capacity/duration constraint is minimized. In the constraints, Equations (1)–(4) and (6) are used. Table 2 shows the solution results.*

$$\min \sum_{k=1}^{r} \sum_{j=1}^{m} (u_{kj}^- + u_{kj}^+)$$

Table 2. Created schedule for Scenario 1 as a result of the goal programming solution.

Day	Block	OR-1	UR (%)	OR-2	UR (%)	OR-3	UR (%)	OR-4	UR (%)	OR-5	UR (%)	OR-6	UR (%)	OR-7	UR (%)	OR-8	UR (%)
1	1	O	100%	GS1	96%	-	-	GS1	84%	CS	90%	O	77%	PS	98%	-	-
	2	CS	79%	O	75%	GS1	84%	PS	58%	GS2	59%	-	-	-	-	PS	97%
2	3	-	-	GS2	97%	GS1	86%	O	73%	PS	98%	CS	31%	-	-	O	96%
	4	O	94%	GS1	98%	PS	77%	GS1	100%	O	90%	-	-	-	-	U	100%
3	5	GS1	82%	U	100%	-	-	CS	100%	GS1	79%	GS2	88%	-	-	O	50%
	6	GS2	89%	-	-	GS2	73%	CS	88%	GS2	100%	U	100%	-	-	CS	100%
4	7	CS	100%	O	96%	GS1	63%	-	-	CS	84%	-	-	PS	87%	PS	82%
	8	-	-	GS1	73%	GS2	90%	O	100%	O	77%	CS	84%	GS1	50%	-	-
5	9	GS1	90%	GS2	100%	GS2	86%	GS2	99%	-	-	GS1	88%	-	-	GS2	90%
	10	GS1	71%	PS	98%	GS1	94%	U	92%	U	79%	-	-	-	-	O	85%

OR: Operating Room, UR: Utilization Rate, O: Orthopedics, CS: Cardiovascular Surgery, GS1: General Surgery 1, GS2: General Surgery 2, U: Urology, PS: Plastic Surgery.

Scenario 2. *In this scenario, block scheduling was performed. Specialists were given the opportunity to choose certain times. According to experts who want to create this scenario, it is thought that the possibility of making a choice is beneficial for surgeons to prepare their own schedules. In this scenario, an orthopedic surgeon is needed to work in each operating room at the time period from 8:00 to 12:00. In mathematical modeling, the orthopedic surgeon's undesirable time zones are outlined. The right side is equal to zero, and deviation in the positive direction in this goal was minimized. This was made to prevent assignment in undesirable time zones.*

In this scenario, the objective function was primarily to minimize deviation in the positive direction in the time-preference constraints of surgeons with certain specialties, deviation in the positive and negative direction in the constraint on the balanced distribution of operating rooms, and deviation in the positive and negative direction at the constraint on using operating room capacity/duration. In the constraints, Equations (1)–(4) and (6)–(8) were used. Table 3 shows the solution results.

$$\min P_1 \sum_{s=1}^{S} r_s^+$$
$$\min P_2 \sum_{j=1}^{m} k_j^- + k_j^+$$
$$\min P_3 \sum_{k=1}^{r} \sum_{j=1}^{m} (u_{kj}^- + u_{kj}^+)$$

Table 3. Created schedule for Scenario 2 as a result of the goal programming solution.

Day	Block	OR-1	UR (%)	OR-2	UR (%)	OR-3	UR (%)	OR-4	UR (%)	OR-5	UR (%)	OR-6	UR (%)	OR-7	UR (%)	OR-8	UR (%)
1	1	O	105%	-	-	PS	100%	-	-	GS2	94%	-	-	GS2	100%	U	70%
	2	GS2	100%	GS2	100%	GS2	90%	GS1	100%	-	-	CS	100%	-	-	CS	100%
2	3	O	75%	PS	105%	-	-	PS	110%	GS1	100%	GS1	73%	-	-	-	-
	4	-	-	GS1	110%	CS	100%	-	-	GS1	100%	-	-	O	100%	GS1	110%
3	5	GS1	100%	-	-	CS	110%	GS2	100%	PS	115%	-	-	U	120%	-	-
	6	-	-	GS2	90%	-	-	GS1	95%	-	-	O	120%	U	110%	-	-
4	7	CS	100%	-	-	O	100%	-	-	PS	115%	GS2	94%	-	-	GS2	100%
	8	-	-	CS	100%	O	85%	CS	110%	-	-	-	-	-	-	GS1	65%
5	9	-	-	O	105%	-	-	GS1	110%	O	120%	GS1	100%	PS	110%	-	-
	10	GS1	100%	-	-	-	-	-	-	-	-	GS2	70%	O	100%	U	110%

OR: Operating Room, UR: Utilization Rate, O: Orthopedics, CS: Cardiovascular Surgery, GS1: General Surgery 1, GS2: General Surgery 2, U: Urology, PS: Plastic Surgery.

Scenario 3. *This scenario studies an expertise in the same operating room for the entire week. With this scenario, the same expertise was assigned throughout the whole time zone of an operating room. Thus, operations of the same expertise could be performed without changing equipment and technical personnel.*

In this scenario, the objective function was primarily to minimize deviation in the positive direction of the constraint of operating rooms belonging to the same specialty, and deviation in the positive and negative direction in the constraint of using operating room capacity/duration. In the constraints, Equations (1)–(6) and (9) were used. Table 4 shows the solution results.

$$\min P_1 \sum_{j=1}^{m} p_j^+$$
$$\min P_2 \sum_{k=1}^{r} \sum_{j=1}^{m} (u_{kj}^- + u_{kj}^+)$$

Table 4. Created schedule for Scenario 3 as a result of the goal programming solution.

Day	Block	OR-1	UR (%)	OR-2	UR (%)	OR-3	UR (%)	OR-4	UR (%)	OR-5	UR (%)	OR-6	UR (%)	OR-7	UR (%)	OR-8	UR (%)
1	1		59%		88%		-		98%		-		88%		86%		29%
	2		84%		65%		-		96%		-		81%		86%		75%
2	3		81%		92%		-		100%		29%		100%		96%		-
	4	GS2	-	CS	71%	U	52%	GS1	96%	GS2	42%	PS	92%	O	86%	GS2	-
3	5		-		96%		96%		98%		-		75%		96%		-
	6		46%		-		48%		96%		-		100%		71%		69%
4	7		-		67%		65%		94%		54%		98%		98%		-
	8		100%		-		75%		94%		73%		100%		100%		-
5	9		-		90%		75%		100%		-		92%		88%		84%
	10		-		94%		84%		92%		-		94%		96%		92%

OR: Operating Room, UR: Utilization Rate, O: Orthopedics, CS: Cardiovascular Surgery, GS1: General Surgery 1, GS2: General Surgery 2, U: Urology, PS: Plastic Surgery.

2.2. Creating a Surgical Schedule in Operating Rooms with Constraint Programming

The specified goals and assumptions under the heading goal programming method were also valid during this solution phase. According to the obtained data, operating room capacities and resource utilization were considered (see Table 1). In this phase, three scenarios were created by considering different situations that were modeled by constraint programming. The time zone and operating room index were converted into single indices, and 80 (= 8 × 10) operating room time periods were named. Within the framework of these assumptions and objectives, every model was coded and resolved with the CP optimizer program.

Notation of model-forming expressions

Decision variables and parameters:

$X_i = i$. operation assignment status $\quad\forall i$

$Y_{ij} = \begin{cases} 1, \text{If operation i is performed in operation room j} \\ 0, \text{otherwise} \end{cases}$ $\quad\forall i, \forall j$

$Z_{js} = \begin{cases} 1, \text{If expert s works in operating room j} \\ 0, \text{otherwise} \end{cases}$ $\quad\forall j, \forall s$

$V_{ts} = \begin{cases} 1, \text{If expert s works in group t} \\ 0, \text{otherwise} \end{cases}$ $\quad\forall t, \forall s$

p_i: operation time i, including cleaning and preparation times $\quad\forall i$
U_j : Daily use time of operating room j $\quad\forall j$
d_s : set from experts s, who prefer to be assigned to the time zone $\quad\forall s$
M: A large number

Deviation variables:

u_j^- : amount of negative deviation to the total available time from operating room time j; $\quad\forall j$
u_j^+ : amount of positive deviation to the total available time from operating room time j; $\quad\forall j$
r_s^- : amount of negative deviation from the preferred time zone by experts s; $\quad\forall s$
r_s^- : amount of negative deviation from the preferred time zone by experts s; $\quad\forall j$
p_t^- : amount of negative deviation from assigned experts in operating room time j; $\quad\forall t$
p_t^+ : amount of positive deviation from assigned experts in operating room time j. $\quad\forall t$

Constraints and goals:

$$x_i \leq 80 \, ; \, x_i \geq 1 \quad \forall i \tag{10}$$

$$x_i \neq x_k \quad \forall i \tag{11}$$

$$(x_i = j) = y_{ij} \quad \forall i, \forall j \tag{12}$$

$$\sum_i y_{ij} \leq M * z_{js} \quad \forall j, \forall s \tag{13}$$

$$\sum_i Z_{js} \leq M * V_{ts} \quad \forall t, \forall s \tag{14}$$

$$\sum_i (p_i * (x_i = j)) + u_j^- - u_j^+ = 240 \quad \forall j \tag{15}$$

$$\sum_{c \in d_s} (x_i = k) + r_j^- - r_j^+ = 0 \quad \forall j \tag{16}$$

$$\sum_s V_{ts} + p_t^- - p_t^+ = 1; \quad \forall t \tag{17}$$

Constraint 10 states that each operation is assigned to the whole operating room and the time zones only once, with operating-time boundaries to which the i operation can be assigned. Constraint 11 ensures that we have, at most, six different specialties in all operating rooms at any time zone. Thus, operations can be carried out with the same equipment and technical personnel in the time zone. An index, such as k, is added at the constraint. While this index specifies the bounds of the number of operations for the relevant expertise in index i in block scheduling, the number of the remaining operations in index k is retained. The constraint for the number of operations of each expertise is written, the number of operations of the previous expertise is also subtracted at each time, and the remaining number of operations is collected in index k. Constraints 12–14 could assign an expert in all operating rooms throughout the entire time zone. Thus, in operating rooms separated by the same expert, operations could be carried out without any wait or interruption due to equipment and technical personnel changes throughout the whole week. Constraint 12, if operation i is assigned within operating room time j, has this value stored in the Y_{ij} decision variable with the corresponding i and j index values. In Constraints 13 and 14, operations are grouped according to their expertise and then blocked with the t-index according to the operating-time group. Thus, an expertise is only assigned to the operating room time group on a column basis.

Constraint 15 aims to minimize operating room overtime and less use of the operating room. It ensures to minimize deviations between reachable time and working time in this goal. The purpose function of this goal is as follows:

$$\min \sum_{j=1}^{m}(u_j^- + u_j^+)$$

Goal Constraint 16 aims at performing a specialist surgeon's operations in each operating room at the desired time zones. When the constraint is run, it is equal to zero by outlining the time zones during which one does not want to be assigned. The c index is expressed as a set of specializations that have preferences. In this goal constraint, deviation in the positive direction is minimized. Thus, attempts are made to prevent assignment to undesired time zones. The purpose function of this goal is as follows:

$$\min \sum_{j=1}^{m} r_j^+$$

Goal Constraint 17 is aimed at allocating each operating room to different specialties during the week. Thus, a specialist works all the time during the week. In this goal constraint, deviation in the positive direction is minimized. The purpose function of this goal is as follows:

$$\min \sum_{t=1}^{T} p_t^+$$

With this mathematical model, three different scenarios were developed. Block scheduling was considered for 120 operations. Block schedules act according to the logic of working certain specialties into specific time zones. These created scenarios are as follows:

Scenario 4. *In this scenario, block scheduling was worked into this model. In this model, there was an attempt to balance operating room use. In order to prevent overtime and less use, same-expertise operations were assigned to the same time zones. In this scenario, the objective function was to minimize deviation in the negative and positive directions related to using operating room capacity/duration constraint. In the constraints, Equations (10), (11) and (15) were used. Table 5 shows the solution results.*

$$\min \sum_{j=1}^{m}(u_j^- + u_j^+)$$

Table 5. Created schedule for Scenario 4 as a result of the constraint programming solution.

Day	Block	OR-1	UR (%)	OR-2	UR (%)	OR-3	UR (%)	OR-4	UR (%)	OR-5	UR (%)	OR-6	UR (%)	OR-7	UR (%)	OR-8	UR (%)
1	1	CS	106%	GS2	59%	O	36%	O	98%	O	90%	O	87%	CS	63%	GS2	50%
	2	O	81%	O	90%	GS2	73%	PS	46%	GS1	71%	PS	75%	CS	32%	GS2	88%
2	3	U	50%	GS1	61%	U	86%	PS	83%	GS1	61%	GS1	88%	GS2	98%	PS	80%
	4	CS	100%	GS1	50%	PS	71%	GS1	52%	PS	40%	CS	40%	GS2	84%	GS2	50%
3	5	PS	57%	U	71%	U	50%	GS2	56%	GS1	98%	U	42%	-	-	PS	50%
	6	PS	40%	GS2	84%	GS2	76%	CS	84%	GS1	25%	GS1	84%	U	61%	GS1	40%
4	7	U	36%	PS	63%	GS1	84%	CS	61%	U	33%	GS1	30%	GS1	82%	O	50%
	8	GS1	48%	CS	77%	GS2	52%	O	48%	O	80%	O	102%	O	84%	U	44%
5	9	PS	48%	GS1	27%	GS1	106%	O	50%	GS1	50%	GS2	86%	CS	104%	GS1	59%
	10	GS2	73%	O	42%	O	100%	PS	44%	GS1	40%	CS	98%	GS2	42%	O	61%

OR: Operating Room, UR: Utilization Rate, O: Orthopedics, CS: Cardiovascular Surgery, GS1: General Surgery 1, GS2: General Surgery 2, U: Urology, PS: Plastic Surgery.

Scenario 5. *In this scenario, block scheduling was performed. Specialists were given the opportunity to choose certain times. In this created scenario, an orthopedic surgeon wanted to work in each operating room between 8:00 and 12:00. At the same time, it ensures that overall expertise distribution is balanced. In the model, the orthopedic surgeon's undesirable time zones are written. The right side is equal to zero, and deviation in the positive direction in this goal was minimized. It was made to prevent assignment to undesirable time zones.*

In this scenario, the objective function was primarily to minimize deviation in the positive direction of the time-preference constraints of surgeons with certain specialties, and deviation in the positive and negative direction of the constraint of using operating room capacity/duration. In the constraints, Equations (10), (11), (15) and (16) were used. Table 6 shows the solution results.

$$\min P_1 \sum_{j=1}^{m} r_j^+$$

$$\min P_2 \sum_{j=1}^{m} (u_j^- + u_j^+)$$

Table 6. Created schedule for Scenario 5 as a result of the constraint programming solution.

Day.	Block	OR-1	UR (%)	OR-2	UR (%)	OR-3	UR (%)	OR-4	UR (%)	OR-5	UR (%)	OR-6	UR (%)	OR-7	UR (%)	OR-8	UR (%)
1	1	O	63%	CS	63%	GS1	88%	U	81%	GS2	46%	CS	46%	PS	86%	GS2	48%
	2	GS2	79%	GS2	82%	PS	50%	PS	38%	-	-	PS	90%	CS	100%	O	55%
2	3	O	48%	O	54%	GS2	46%	O	50%	GS2	50%	PS	59%	GS1	48%	PS	46%
	4	U	50%	GS1	50%	O	109%	GS1	46%	GS2	46%	O	98%	PS	69%	GS1	81%
3	5	GS1	68%	O	61%	CS	92%	GS2	96%	GS1	84%	PS	69%	GS1	49%	GS2	44%
	6	PS	88%	U	59%	O	46%	O	50%	O	92%	GS2	59%	CS	100%	O	56%
4	7	PS	73%	GS2	69%	CS	96%	GS1	71%	GS1	84%	O	94%	GS2	46%	O	54%
	8	GS1	61%	GS1	100%	CS	94%	CS	84%	GS1	61%	GS2	50%	U	71%	GS1	96%
5	9	O	50%	PS	64%	GS1	84%	GC1	48%	GS2	38%	GS2	98%	U	61%	CS	100%
	10	GS1	123%	U	59%	U	50%	U	73%	GS1	38%	GS2	93%	U	50%	GS2	46%

OR: Operating Room, UR: Utilization Rate, O: Orthopedics, CS: Cardiovascular Surgery, GS1: General Surgery 1, GS2: General Surgery 2, U: Urology, PS: Plastic Surgery.

Scenario 6. *This scenario studies a type of expertise in the same operating room for the entire week. In this scenario, which is more desirable to realize in real life, the same type of expertise was assigned throughout the whole time zone of an operating room. Thus, operations of the same expertise could be performed without changing equipment and technical personnel.*

In this scenario, the objective function is primarily to minimize deviation in the positive direction of the constraint of operating rooms belonging to the same specialty, and deviation in the positive and negative direction of the constraint of using operating room capacity/duration. In the constraints, Equations (10)–(15) and (17) are used. Table 7 shows the solution results.

$$min P_1 \sum_{t=1}^{T} p_t^+$$

$$min P_2 \sum_{j=1}^{m} (u_j^- + u_j^+)$$

Table 7. Created schedule for Scenario 6 as a result of the constraint programming solution.

Day	Block	OR-1	UR (%)	OR-2	UR (%)	OR-3	UR (%)	OR-4	UR (%)	OR-5	UR (%)	OR-6	UR (%)	OR-7	UR (%)	OR-8	UR (%)
1	1	O	121%	GS2	88%	U	40%	CS	63%	GS1	28%	PS	113%	KD	-	GS1	50%
	2		67%		110%		61%		63%		88%		111%		-		-
2	3		121%		92%		94%		-		84%		34%		-		50%
	4		98%		50%		34%		-		67%		60%		100%		84%
3	5		102%		86%		46%		82%		80%		57%		84%		84%
	6		117%		111%		-		-		88%		38%		25%		42%
4	7		104%		84%		-		-		20%		38%		75%		71%
	8		84%		150%		86%		-		110%		65%		34%		61%
5	9		110%		90%		42%		100%		130%		98%		67%		73%
	10		92%		63%		71%		75%		25%		88%		-		-

OR: Operating Room, UR: Utilization Rate, O: Orthopedics, CS: Cardiovascular Surgery, GS1: General Surgery 1, GS2: General Surgery 2, U: Urology, PS: Plastic Surgery.

3. Results

In this study, the integrated-goal programming and constraint programming methods were used in the operating room scheduling processes by using data from a state hospital. Both methods have decision variables, an objective function, and restrictive equations. Table 8 summarizes the current situation and scenarios.

Table 8. Summary of current situation and scenarios.

	Current Situation	Scenario 1 & 4		Scenario 2 & 5		Scenario 3 & 6	
		GP	CP	GP	CP	GP	CP
Number of Overtime Time-Zone	45	-	4	17	2	-	13
Number of Less Use Time-Zone (Under 50%)	15	-	20	-	16	5	10
Number of Full Capacity Time-Zone	5	11	2	20	4	7	2
Number of Unused Time-Zone	-	20	1	32	1	21	13
Total Utilization Rate	52%	85%	70%	95%	70%	80%	79%
Preferred status	-	-	-	√	√	-	-
Closing the operating room	-	-	-	-	-	√	√
Block planning strategy	√	√	√	√	√	√	√
Balanced assignment	-	√	√	√	√	√	√

GP: Goal Programming, CP: Constraint Programming.

Table 8 shows the success rates of both methods, modeled with integrated goal programming and constraint programming, in these scenarios. Looking at the detailed schedules, Scenarios 1 and 4 have all operations assigned through the goal programming method. The 11 time zones in the operating rooms have a full-capacity utilization rate. Thus, in some time zones, operating rooms are closed and prevented from idly waiting. Surgeons' working hours are also programmed during working hours without overtime. Costs caused by overtime were avoided, and surgeons' satisfaction was increased. In constraint programming, simplicity is provided in the modeling processes. In constraint programming, only the seventh operating room does not work on the third day between 08:00 and 12:00. All patient operations on the waiting list are carried out. This increased the level of patient satisfaction. However, in the 20 time zones, capacity was below the desired utilization level. However, according to the solution results, the utilization rates of the operating rooms were decreased. This situation is likely to result in costs that arise from operating room waiting times. Scenarios 2 and 5: all operations were assigned with the goal programming method. All operating room utilization rates were intended to use full capacity, but it seems that overtime was done in 17 time zones. In this case, surgeon satisfaction levels may be reduced and, at the same time, the desired surgery efficiency may be affected. It could also cause overtime costs to increase. In constraint programming, there is a capacity utilization rate at the desired level in operating rooms. In this method, overtime was only observed in two time zones, and all waiting-list patient operations were performed. The level of satisfaction of both surgeons and patients increased without overtime. In addition, overtime costs were avoided. The orthopedic surgeon wanting to work during the desired time intervals was achieved in both methods. Scenarios 3 and 6: the operating rates of operating rooms were close to 100% and worked without overtime in goal programming. Overtime was observed in operating rooms where the orthopedic surgeon appointed the constraint program. This suggests that the model is difficult to distribute over time. According to the results of the two methods, the schedule showed that the total utilization rates were very close to each other.

Looking at Table 8, it can be seen that the obtained results are directed towards the efficiency that hospital managers want to obtain from operating rooms. Results were achieved with very little deviation. The established mathematical models were run for 3000 s. Concerning model deviations, in the results of Scenarios 1and 4, there were no deviations that would cause overtime in the goal programming results. In the case of constraint programming, a deviation of over 60 min was observed. In the results of Scenarios 2 and 5, there were 255 min of overtime in the goal programming method, and 70 min of overtime in the constraint programming method. In Scenarios 3 and 6, there was no overtime in the goal programming method. In the constraint programming method, 125 min overtime occurred. Deviation values indicate the overtime time intervals in the operating rooms. The reason for overtime in these scenarios was due to special constraints in the models. When hospital administrators provide opportunities such as special preference for surgeons and the separation of special operations of operating rooms, it can be seen that overtime occurred in the model. However, when the results are examined, the best results are obtained in the time constraint. Overtime work is done in operating rooms as little as possible, and the desired constraints and other purposes are provided. In the model, there was no deviation in other purpose constraints in the objective function. In addition, the number of decision variables in the constraint programming method was less than the number of decision variables in the goal programming model. This facilitated the establishment and resolution of the model. The results show that the operating rooms in the hospital were used as effectively as possible. In other words, it is understood that overtime rates, waiting periods, and the unnecessary use of resources in hospitals could be reduced. Reducing these conditions increases operating room efficiency. Operations can be carried out in the current time, which is separated by decreased waiting and postponements. Resources can be effectively used. At the same time, these results allow the study to be reprogrammable. In this case, it is possible to maximize the number of patients who can be treated with idle block times. Thus, it is possible to increase the obtained performance and efficiency from operating rooms. The fact that the study is open to development in this respect shows that it may have a fundamental nature. The fact that this study is adaptable according to the hospital structure emphasizes the preferable feature of hospital managers. The purpose of reducing the desired costs of operating rooms and increasing their efficiency is the overlap of the obtained results from this study. In the literature [36–38], the main purpose of operating rooms is to reduce costs. These costs are mostly due to overtime and waiting. In order to reduce these costs, researchers benefit from different methods. Thus, they obtain high-quality and efficient programs with the methods they use in operating room scheduling processes [39]. The aim of this study was to prevent confusion and possible problems caused by the mixed performance of hospital operations. In other words, with the assignment of different types of operations in succession to the same operating room, it is desired to increase efficiency from the equipment and surgical team. In these results, these situations are prevented. At the same time, the impact of these situations is reduced and performance is improved, allowing for a more robust and flexible start-up program.

In the current case, since schedules are manually prepared, not all operations could be transferred to the schedules, which led to an increase in patient waiting times. With these created schedules, all operations were assigned. Even though the created schedules had an overtime status in the time zones, the operating rooms were used with full capacity. In the current case, the utilization of operating rooms was not effective because the assignment of operations could not be done in a balanced way.

4. Discussion and Conclusion

The inability of surgeons to promptly enter operations, which is one of the problems experienced in hospitals today, causes both the inability to effectively use existing hospital capacity, and the decrease of offered service quality to patients. In this study, a flexible model proposal is presented with integrated goal programming and constraint programming methods. It aims to increase patient/staff satisfaction by ensuring the efficient use of hospital resources. Three different scenarios were created for the purposes of hospital managers. According to these scenarios, the established models were run for 3000 s. According to the appropriate solutions, obtained as a result of 3000 s, scenario charts were created. In the first scenario, on the basis of the block-scheduling strategy, the aim was the balanced distribution of operations in operating rooms and the effective use of operating rooms. In the second scenario, in addition to these goals, an objective for the orthopedic surgeon to only work during morning hours was determined, and the balanced distribution of operations in the time zones was desired. In the third scenario, it was desirable that an operating room be divided into only one specialty during all time zones. All these scenarios are modeled by the integrated-goal programming and constraint programming methods. Then, the method results were compared and analyzed according to the current situation.

When creating the model, we tried to create weekly schedules, together with the time-zone index, without complicating the problem structure. In the scenarios, utilization rates were a measure of how effectively operating room capacity was being used. This shows that the methods help to make the process systematic. According to the current situation, schedule efficiency was increased with the created models, and the best possible utilization of the operating room was achieved. It is aimed that this study will be an example for future studies. With the integration of constraint programming and goal programming, it has contributed to the literature by increasing schedule effectiveness. With this integration, scenarios were created for different desired goals from other work.

When other studies in the literature are examined, it is seen that operating room scheduling studies have frequently been discussed by the researchers in the last forty years [39]. Studies [39–42] indicate that operating room planning is affected by personnel and equipment-based factors. Therefore, multiple objectives cannot be taken into consideration in planning and scheduling. In this context, problems can be solved by developing strategies for solution techniques. Researchers use various solution methods in operating room scheduling problems [43–45]. Each method has its own advantage. These methods are also determined according to the problem type. In this study, unlike other studies, the integrated goal programming and constraint programming methods, which are effective for the simultaneous realization of the desired objectives, were used. This integration allows flexibility in the objectives and enables to produce solutions to real-life problems [33]. At the same time, in this study, the realization of situations encountered in real-life scenarios differentiates the process from other studies. In fact, multiple real-life problems could be considered at the same time with this scenario. In this case, a systematic structure was developed to solve problems encountered in hospitals. In addition, the integrated goal programming and constraint programming methods enable the solution process to be suitable for real life by operationally strengthening the study. Because the work allows flexibility, it is easy to create a new start-up program at times when reprogramming is needed. The discussed multipurpose problem was solved for the single purpose of modelling. The objectives to be carried out in the solution process were combined with the goal programming method, and the problem was modelled using the constraint programming method. With the systematic structure established, the model can quickly be reviewed without major and complex changes. In this context, this study comes to the fore in the literature. In this study, the flexibility and advantages of goal programming and constraint programming were utilized. The constraint programming method facilitates writing logical constraints in the modelling process. At the same time, decision variables can take values in different structures besides numerical variables. In the solution processes, the constraint programming method allows decision-makers to establish models with fewer decision variables. Decision makers can use the search algorithms defined in the modelling phase. Considering the advantages of the constraint programming method, the basic constraint programming model is proposed in the problem solution. The operating room scheduling problem mainly deals with which operation should be assigned to which operating room. In this problem, the decision variable structure is a 0–1 logic structure. If the operation is assigned to the operating room, it takes the value of 1; if it is not assigned, it takes 0. Different solution approaches were developed by researchers for the scheduling problems involving this 0–1 variable structure. The constraint programming method also allows to model variables in the 0–1 structure as an integer [46]. The constraint programming method has a structure that can work more efficiently with relatively few variables in the modelling. With this feature, constraint programming provides better performance by reducing the domain. This problem is translated from a 0–1 structure into an integer structure by taking advantage of the modelling power of constraint programming [46].

At the same time, in the literature, researchers mostly focus on an open-planning strategy as they facilitate the solution process. However, this situation causes operations with the same expertise in operating rooms to be assigned to different operating rooms at different times. This necessitates the permanent displacement of equipment used during operations in operating rooms. Most of the time, due to these displacements, equipment damage or patient/surgeon waiting times increase. In this context, all these reasons exponentially cause negative consequences to hospital managers. Considering all this, a block-scheduling strategy was adopted in this study to avoid negative results, and to ensure patient/surgeon satisfaction in the hospital. When block-scheduling studies [11,15,16,20,47–49] in the literature are examined, it is understood that they want to avoid these negative results.

In future studies, the inclusion of different constraints could be considered with regard to the special circumstances of surgeons' working conditions. Depending on each surgeon's preferences, prioritization and assignments can be made using multicriteria decision-making methods in operations. Various situations could be considered, such as the fact that special circumstances related to the staff are reflected in the constraints.

5. Limitations

The main limitation of this study is that the available data are composed of historical data. It is very difficult for researchers to plan for current real-life situations. There are too many restrictions to consider. It is not possible to solve all these limitations in a model at the same time. Therefore, certain assumptions were made in this study.

Author Contributions: Ş.G. conceptualized the study, prepared dataset, conducted analyses, contributed to writing the manuscript, and provided modeling processes. T.E. provided overall guidance and expertise in conducting the analysis and was a contributor to writing the manuscript. H.M.A. supported the solution process of the mathematical model and was a contributor to writing the manuscript. All authors read and approved the final manuscript.

Funding: This research was funded by Scientific Research Program (BAP) of Kirikkale University as project of 2017/027.

Conflicts of Interest: The authors declare no conflict of interest.

Ethics Approval and Consent to Participate: This study does not require ethical approval.

References

1. Pham, D.N.; Klinkert, A. Surgical Case Scheduling as A Generalized Job Shop Scheduling Problem. *Eur. J. Oper. Res.* **2008**, *185*, 1011–1025. [CrossRef]
2. Baesler, F.; Gatica, J.; Correa, R. Simulation Optimization for Operating Room Scheduling. *Int. J. Simul. Model.* **2015**, *14*, 215–226. [CrossRef]
3. Özkarahan, I. Allocation of Surgeries to Operating Rooms by Goal Programing. *J. Med. Syst.* **2000**, *24*, 339–378. [CrossRef] [PubMed]
4. Arenas, M.; Bilbao, A.; Caballero, R.; Gómez, T.; Rodríguez, M.V.; Ruiz, F. Analysis Via Goal Programming of the Minimum Achievable Stay in Surgical Waiting Lists. *J. Oper. Res. Soc.* **2002**, *53*, 387–396. [CrossRef]
5. Blake, J.T.; Donald, J. Mount Sinai Hospital Uses Integer Programming to Allocate Operating Room Time. *Interfaces* **2002**, *32*, 63–73. [CrossRef]
6. Blake, J.T.; Carter, M.W. Goal Programming Approach to Strategic Resource Allocation in Acute Care Hospitals. *Eur. J. Oper. Res.* **2002**, *140*, 541–561. [CrossRef]
7. Kharraja, S.; Albert, P.; Chaabane, S. Block Scheduling: Toward A Master Surgical Schedule. In Proceedings of the 2006 International Conference on Service Systems and Service Management, Troyes, France, 25–27 October 2006; pp. 429–435.
8. Wullink, G.; Van Houdenhoven, M.; Hans, E.W.; van Oostrum, J.M.; van der Lans, M.; Kazemier, G. Closing Emergency Operating Rooms Improves Efficiency. *J. Med. Syst.* **2007**, *31*, 543–546. [CrossRef] [PubMed]
9. Paoletti, X.; Marty, J. Consequences of Running More Operating Theatres Than Anaesthetists To Staff Them: A Stochastic Simulation Study. *Br. J. Anaesth.* **2007**, *98*, 462–469. [CrossRef] [PubMed]

10. Lamiri, M.; Xie, X.; Dolgui, A.; Grimaud, F. A Stochastic Model for Operating Room Planning with Elective and Emergency Demand for Surgery. *Eur. J. Oper. Res.* **2008**, *185*, 1026–1037. [CrossRef]
11. Beliën, J.; Demeulemeester, E.; Cardoen, B. A Decision Support System for Cyclic Master Surgery Scheduling with Multiple Objectives. *J. Sched.* **2009**, *12*, 147–161. [CrossRef]
12. Zhang, B.; Murali, P.; Dessouky, M.M.; Belson, D. A Mixed Integer Programming Approach for Allocating Operating Room Capacity. *J. Oper. Res. Soc.* **2009**, *60*, 663–673. [CrossRef]
13. Çekic, B. Scheduling of Operating Room Activities and a Hospital Practice. Master's Thesis, Institute of Social Sciences, Hacettepe University, Ankara, Turkey, 2006.
14. Fei, H.; Meskens, N.; Chu, C. A Planning and Scheduling Problem for An Operating Theatre Using an Open Scheduling Strategy. *Comput. Ind. Eng.* **2010**, *58*, 221–230. [CrossRef]
15. Tànfanı, E.; Testi, A. A Pre-Assignment Heuristic Algorithm for the Master Surgical Schedule Problem (MSSP). *Ann. Oper. Res.* **2010**, *178*, 105–119. [CrossRef]
16. Banditori, C.; Cappanera, P.; Visintin, F. A Combined Optimization–Simulation Approach to the Master Surgical Scheduling Problem. *Ima J. Manag. Math.* **2013**, *24*, 155–187. [CrossRef]
17. Cappanera, P.; Visintin, F.; Banditori, C. Comparing Resource Balancing Criteria in Master Surgical Scheduling: A Combined Optimization-Simulation Approach. *Int. J. Prod. Econ.* **2014**, *158*, 179–196. [CrossRef]
18. Eren, T.; Kodanlı, E.; Altundag, B.; Salim, K.M.; Sultan, Ü.; İsmail, B.; Kenan, T. Operating Room Scheduling and A Case Study (In Turkish). *J. Econ. Bus. Politics Int. Relat.* **2016**, *2*, 71–85.
19. Xıang, W. A Multi-Objective ACO For Operating Room Scheduling Optimization. *Nat. Comput.* **2017**, *16*, 607–617. [CrossRef]
20. Abedini, A.; Li, W.; Ye, H. An Optimization Model for Operating Room Scheduling to Reduce Blocking Across the Perioperative Process. *Procedia Manuf.* **2017**, *10*, 60–70. [CrossRef]
21. Dagdeviren, M.; Eren, T. Analytical Hierarchy Process and Use Of 0-1 Goal Programming Methods in Selecting Supplier Firm. *Gazi Univ. J. Eng. Archit.* **2001**, *16*, 41–52.
22. Demirtas, E.A.; Ustun, Ö. Analytic Network Process and Goal Programming Approach in Supplier Selection and Order Allocation. In Proceedings of the Operations Research/Industrial Engineering—XXIV National Congress, Gaziantep, Turkey, 15–18 June 2004.
23. Özder, E.H.; Eren, T.; Çetin Özel, S. Supplier Selection with TOPSIS and Goal Programming Methods: A Case Study. *J. Trends Dev. Mach. Assoc. Technol.* **2015**, *19*, 109–112.
24. Özder, E.H.; Eren, T. Supplier Selection with in Multi Criteria Decision Making Method and Goal Programming Techniques. *Selcuk Univ. J. Eng. Sci. Technol.* **2016**, *4*, 196–207.
25. Gür, Ş.; Hamurcu, M.; Eren, T. Using Analytic Network Process and Goal Programming Methods for Project Selection in the Public Institution. *Les Cah. Du Mecas* **2016**, *13*, 36–51.
26. Hamurcu, M.; Gür, Ş.; Özder, E.H.; Eren, T. A Multicriteria Decision Making for Monorail Projects with Analytic Network Process and 0-1 Goal Programming. *Int. J. Adv. Electron. Comput. Sci.* **2016**, *3*, 8–12.
27. Gül, E.; Eren, T. Logistics Distribution Network Problems and Warehouse Selection with Goal Programming and Analytical Hierarchy Process. *Harran Univ. J. Eng.* **2017**, *2*, 1–13.
28. Taha, H.A. *Operations Research: An Introduction*; Macmillan: London, UK, 1992.
29. Varlı, E. Solution of the Shift Scheduling Problem for the Foremans in the Manufacturing Sector with AHP-Goal Programming. Master's Thesis, Institute of Science and Technology, Kirikkale University, Kırıkkale, Turkey, 2017.
30. Focacci, F.; Laburthe, F.; Lodi, A. Local Search and Constraint Programming. In *Handbook of Metaheuristics*; Springer: Berlin, Germany, 2003; pp. 369–403.
31. Apt, K. *Principles of Constraint Programming*; Cambridge University Press: Cambridge, UK, 2003.
32. Alağaş, H.M. Constraint Programming Model and Search Strategies for the Mixed Model Assembly Line Balancing Problem. Ph.D. Thesis, Institute of Science and Technology, Gazi University, Ankara, Turkey, 2017.
33. Gür, Ş.; Eren, T. Scheduling and Planning in Service Systems with Goal Programming: Literature Review. *Mathematics* **2018**, *6*, 265. [CrossRef]
34. Rossi, F.; Van Beek, P.; Walsh, T. (Eds.) *Handbook of Constraint Programming*; Elsevier: Amsterdam, The Netherlands, 2006.

35. IBM ILOG CPLEX. "12.6." CPLEX User's Manual. 2014. Available online: https://www.ibm.com/support/knowledgecenter/SSSA5P_12.6.2/ilog.odms.cplex.help/CPLEX/homepages/usrmancplex.html (accessed on 8 March 2019).
36. Hooshmand, F.; MirHassani, S.A.; Akhavein, A. Adapting GA to solve a novel model for operating room scheduling problem with endogenous uncertainty. *Oper. Res. Health Care* **2018**, *19*, 26–43. [CrossRef]
37. Liu, H.; Zhang, T.; Luo, S.; Xu, D. Operating room scheduling and surgeon assignment problem under surgery durations uncertainty. *Technol. Health Care* **2018**, *26*, 297–304. [CrossRef] [PubMed]
38. Molina-Pariente, J.M.; Hans, E.W.; Framinan, J.M. A stochastic approach for solving the operating room scheduling problem. *Flex. Serv. Manuf. J.* **2018**, *30*, 224–251. [CrossRef]
39. Gür, Ş.; Eren, T. Application of operational research techniques in operating room scheduling problems: Literatüre overview. *J. Healthc. Eng.* **2018**, *2018*, 1–15. [CrossRef] [PubMed]
40. Cayirli, T.; Veral, E. Outpatient scheduling in healthcare: A review of literature. *Prod. Oper. Manag.* **2003**, *12*, 519–549. [CrossRef]
41. Cardoen, B.; Demeulemeester, E.; Beliën, J. Operating room planning and scheduling: A literatüre review. *Eur. J. Oper. Res.* **2010**, *201*, 921–932. [CrossRef]
42. Guerriero, F.; Guido, R. Operational research in the management of the operating theatre: A survey. *Healthc. Manag. Sci.* **2011**, *14*, 89–114. [CrossRef] [PubMed]
43. Augusto, V.; Xie, X.; Perdomo, V. Operating theatre scheduling using Lagrangian relaxation. *Eur. J. Ind. Eng.* **2008**, *2*, 172–189. [CrossRef]
44. Fügener, A.; Hans, E.W.; Kolisch, R.; Kortbeek, N.; Vanberkel, P.T. Master surgery scheduling with consideration of multiple downstream units. *Eur. J. Oper. Res.* **2014**, *239*, 227–236. [CrossRef]
45. Visintin, F.; Cappanera, P.; Banditor, C.; Danese, P. Development and implementation of an operating room scheduling tool: An action research study. *Prod. Plan. Control* **2017**, *28*, 758–775. [CrossRef]
46. Leung, J.Y. (Ed.) *Handbook of Scheduling: Algorithms, Models, and Performance Analysis*; CRC Press: Boca Raton, FL, USA, 2004.
47. Mannino, C.; Nilssen, E.J.; Nordlander, T.E. A pattern based, robust approach to cyclic master surgery scheduling. *J. Sched.* **2012**, *15*, 553–563. [CrossRef]
48. Agnetis, A.; Coppi, A.; Corsini, M.; Dellino, G.; Meloni, C.; Pranzo, M. A decomposition approach for the combined master surgical schedule and surgical case assignment problems. *Health Care Manag. Sci.* **2014**, *17*, 49–59. [CrossRef] [PubMed]
49. Nino, L.; Harris, S.; Claudio, D. A simulation of variability-oriented sequencing rules on block surgical scheduling. In Proceedings of the 2016 Winter Simulation Conference (WSC), Washington, DC, USA, 11–14 December 2016.

 © 2019 by the authors. Licensee MDPI, Basel, Switzerland. This article is an open access article distributed under the terms and conditions of the Creative Commons Attribution (CC BY) license (http://creativecommons.org/licenses/by/4.0/).

Article

Chebyshev Spectral Collocation Method for Population Balance Equation in Crystallization

Chunlei Ruan

School of Mathematics & Statistics, Henan University of Science & Technology, Luoyang 471023, China; ruanchunlei@haust.edu.cn

Received: 18 February 2019; Accepted: 26 March 2019; Published: 28 March 2019

Abstract: The population balance equation (PBE) is the main governing equation for modeling dynamic crystallization behavior. In the view of mathematics, PBE is a convection–reaction equation whose strong hyperbolic property may challenge numerical methods. In order to weaken the hyperbolic property of PBE, a diffusive term was added in this work. Here, the Chebyshev spectral collocation method was introduced to solve the PBE and to achieve accurate crystal size distribution (CSD). Three numerical examples are presented, namely size-independent growth, size-dependent growth in a batch process, and with nucleation, and size-dependent growth in a continuous process. Through comparing the results with the numerical results obtained via the second-order upwind method and the HR-van method, the high accuracy of Chebyshev spectral collocation method was proven. Moreover, the diffusive term is also discussed in three numerical examples. The results show that, in the case of size-independent growth (PBE is a convection equation), the diffusive term should be added, and the coefficient of the diffusive term is recommended as $2G \times 10^{-3}$ to $G \times 10^{-2}$, where G is the crystal growth rate.

Keywords: spectral collocation method; population balance equation; Chebyshev points; crystallization

1. Introduction

Crystallization operations are widely used in chemical and pharmaceutical industries. Crystallization kinetics is the basis for the development of crystallization theory. Simultaneously, studies on crystalline kinetics also guide the design of crystallizers and the quality control of crystalline products [1]. The population balance equation (PBE) is a widely accepted equation to model the dynamic crystallization behavior [2]. Therefore, it is of great significance to accurately solve the PBE and to achieve accurate crystal size distribution (CSD).

Currently, there are a many numerical results on the study of PBE. In terms of numerical methods, they can be roughly divided into the following five categories [1,2]: (1) the method of moments [3]; (2) the method of characteristics [4,5]; (3) the method of weight residuals/orthogonal collocation [6,7]; (4) the Monte Carlo method [8,9]; (5) the finite-difference schemes/discrete population balances [10]. The review by Ramkrishna [2] is helpful in this regard. The method of moments approximates the CSD through its moments. The predictions under certain conditions are closed. However, the moment closure conditions are violated for more complex systems. The method of characteristics tries reducing the partial differential equations (PDEs) to ordinary differential equations (ODEs) using lines of characteristics. The approach is efficient with simple physics, but it is not suitable for complex physics. The method of weight residuals/orthogonal collocation approximates the distribution by the basis function; the main weakness is the dependence on the choice of basis function. The Monte Carlo method tracks the histories of individual particles. It has an incomparable degree of omnipotence. However, its computational cost is relatively expensive. The finite-difference/volume methods are approximated by finite-difference/volume schemes. They can be applied to complex cases. However, they usually require a high number of grid points in order to achieve desirable accuracy.

Recently, a large number of numerical methods with high accuracy were constructed to solve PBE, e.g., the HR-van method with a flux limiter [1,11], the Lattice Boltzmann method [12], the weighted essentially non-oscillatory (WENO) method [13], the spectral method [14,15], etc. Among these, the spectral method has the advantage of being able to accurately solve the PBE using a lower number of grid points. It was also found that the spectral collocation method is more efficient than other spectral methods [14] and finite-volume schemes [15].

Because the PBE is a convection–reaction equation in mathematics, its strong hyperbolic property may challenge the numerical methods. The numerical methods may gain unstable results, which are not applicable for physics. In order to weaken the hyperbolic property of PBE, a diffusive term is added herein. Although the diffusive term makes the numerical solution deviate from the exact solution to some extent, it guarantees the stability of the numerical results.

In this paper, Chebyshev spectral collocation is presented to solve the one-dimensional PBE. The CSD of three types of crystallization in batch crystallizers and continuous crystallizers were calculated. By comparing the analytical results and other numerical results, the validity and high accuracy of the Chebyshev spectral collocation method were verified. Furthermore, the effects of the diffusive term are also discussed using three numerical examples.

2. One-Dimensional PBE and Numerical Schemes

2.1. One-Dimensional PBE

PBE is a differential equation which describes the evolution of a population of particles. In the crystallization process, it can reflect the crystal nucleation, growth, agglomeration, and breakage [2].

One-dimensional PBE [1,2,11] can be written as follows:

$$\frac{\partial f(L,t)}{\partial t} + \frac{\partial \{G(L,t)f(L,t)\}}{\partial L} = h(L,t,f), \quad (1)$$

where $f(L,t)$ is the crystal size distribution function, L is the crystal size variable, t is the time, $G(L,t)$ is the growth rate of crystals, and $h(L,t,f)$ is the source term and may include the aggregation, nucleation, breakage, etc. Here, crystal size variable L is typically the characteristic length, volume, or mass, but it can also represent age, composition, and other characteristics of an entity in a distribution [1]. The growth rate $G(L,t)$ can be a function of size and other variables, such as the temperature and the concentration of chemical species in solution [1]. The source term $h(L,t,f)$ is a creation/depletion rate, and it can be a function of other variables including the distribution, which occurs in nucleation processes resulting from particle–particle interactions and in agglomeration processes [1]. Since many of these processes involve integrals, PBE is considered an integro-differential equation describing the complex crystallization processes [2].

Since Equation (1) is a convection–reaction equation, its strong hyperbolic property may challenge the numerical methods. In order to weaken the hyperbolic property of Equation (1), a diffusive term "$\varepsilon \frac{\partial^2 f(L,t)}{\partial L^2}$" is added to the left in Equation (1), leading to

$$\frac{\partial f(L,t)}{\partial t} + \frac{\partial \{G(L,t)f(L,t)\}}{\partial L} + \varepsilon \frac{\partial^2 f(L,t)}{\partial L^2} = h(L,t,f). \quad (2)$$

Here, ε is a small positive number. Theoretically, the closer ε is to 0, the more accurate the result is. However, in numerical experiments, when ε approaches 0, the effect of the diffusive term becomes smaller, which leads to an unstable numerical solution. Therefore, in this paper, the value of ε was determined by numerical tests.

2.2. Chebyshev Spectral Collocation Method

The Chebyshev spectral collocation method is a kind of collocation method [16,17]. On the one hand, as a spectral method, it has high accuracy. On the other hand, as a collocation method, it transforms the derivative term into a differential matrix, which makes the equations simple and easy to solve. To date, this method was successfully applied to solve differential equations in the fields of fluid mechanics [18,19], quantum mechanics [20,21], etc.

Equation (2) can be written as follows:

$$\frac{\partial f(L,t)}{\partial t} = R(L,t,f), \tag{3}$$

where $R(L,t,f) = h(L,t,f) - \frac{\partial \{G(L,t)f(L,t)\}}{\partial L} - \varepsilon \frac{\partial^2 f(L,t)}{\partial L^2}$. The fourth-order Runge–Kutta method [22] is used to discretize the time, leading to

$$\begin{aligned} f^{n+1} &= f^n + \frac{1}{6}(k_1 + 2k_2 + 2k_3 + k_4) \\ k_1 &= \Delta t \cdot R(t_n, f^n(L)) \\ k_2 &= \Delta t \cdot R(t_n + \frac{\Delta t}{2}, f^n(L) + \frac{1}{2}k_1) \\ k_3 &= \Delta t \cdot R(t_n + \frac{\Delta t}{2}, f^n(L) + \frac{1}{2}k_2) \\ k_4 &= \Delta t \cdot R(t_{n+1}, f^n(L) + k_3) \end{aligned} \tag{4}$$

Here, Δt is the time step, $t_n = n\Delta t$ is the time on the nth step ($n = 0, 1, \cdots$), f^n is the value of f on t_n and the initial distribution is written as f^0. It is worth mentioning that the variable f^n in Equation (4) is only a function of the spatial variable L. In this way, the space can be discretized using the spectral collocation method.

In the spectral collocation method, the solution of $f^n(L)$ can be written as follows [16,17]:

$$f^n(L) = \sum_{k=0}^{N} P_k(L) u_k, \tag{5}$$

where $P_k(L)$ is the Lagrange interpolation polynomial at the node $L_k (k = 0, 1, \cdots N)$, which is

$$P_k(L) = \frac{(L-L_0) \cdots (L-L_{k-1})(L-L_{k+1}) \cdots (L-L_N)}{(L_k-L_0) \cdots (L_k-L_{k-1})(L_k-L_{k+1}) \cdots (L_k-L_N)}, \tag{6}$$

and u_k is the function value at interpolation node L_k.

In the Chebyshev spectral collocation method, Chebyshev points are chosen as the interpolation nodes. In the standard calculation domain $[-1,1]$, the Chebyshev points are $\hat{L}_j = \cos(j\pi/N)$ ($j = 0, 1, \cdots N$). As for the general calculation domain $[a,b]$, the Chebyshev points can be obtained by the coordinate transformation of standard Chebyshev points, namely $L_j = \frac{b-a}{2}\hat{L}_j + \frac{a+b}{2}$.

Here, we take the calculation of $R(t_n, f^n(L))$ in Equation (4) as an example to show the implementation of the Chebyshev spectral collocation method. In fact, the right-hand term $R(t_n, f^n(L))$ is equal to $h(t_n, f^n(L)) - \frac{dG(L)}{dL} f^n(L) - G(L)\frac{df^n(L)}{dL} - \varepsilon \frac{d^2 f^n(L)}{dL^2}$. Here, we need to show the calculation of $\dfrac{df^n(L)}{dL}$ and $\dfrac{d^2 f^n(L)}{dL^2}$ terms. In the Chebyshev spectral collocation method [16,17], the derivative term $\dfrac{df^n(L)}{dL}$ can be transformed into a differential matrix as follows:

$$\dfrac{df^n(L)}{dL} = D_N u. \tag{7}$$

Here, D_N is a $(N+1) \times (N+1)$ Chebyshev derivative matrix, and $u = (u_0, u_1, \cdots, u_N)^T$ is a $N+1$ function value vector at the interpolation nodes. In the standard calculation domain $[-1,1]$, the Chebyshev derivative matrix $\hat{D}_N$ satisfies the following properties [16,17]:

$$\begin{aligned} (\hat{D}_N)_{00} &= \dfrac{2N^2+1}{6}, (\hat{D}_N)_{NN} = -\dfrac{2N^2+1}{6} \\ (\hat{D}_N)_{jj} &= \dfrac{-x_j}{2(1-x_j^2)}, j = 1, \cdots, N-1 \\ (\hat{D}_N)_{ij} &= \dfrac{c_i(-1)^{i+j}}{c_j(x_i - x_j)}, i \neq j, i,j = 0, \cdots, N \end{aligned} \tag{8}$$

where $(\hat{D}_N)_{ij}$ represents the element of the $i+1$th row and $j+1$th column in the Chebyshev matrix, and c_i is the constant which is determined by the following equation:

$$c_i = \begin{cases} 2, & i = 0, N \\ 1, & i = 1, \cdots, N-1 \end{cases}. \tag{9}$$

As for the general calculation domain $[a,b]$, the Chebyshev derivative matrix D_N can be calculated using the standard Chebyshev derivative matrix $\hat{D}_N$, which is

$$D_N = \dfrac{2}{a+b}\hat{D}_N. \tag{10}$$

Similarly, in the Chebyshev spectral collocation method [16,17], the second-order derivative term $\dfrac{d^2 f^n(L)}{dL^2}$ can be written as follows:

$$\dfrac{d^2 f^n(L)}{dL^2} = D_N^2 u. \tag{11}$$

Here, D_N^2 is the second-order Chebyshev derivative matrix. The relationship between the second-order Chebyshev derivative matrix D_N^2 and the first-order Chebyshev derivative matrix D_N is

$$D_N^2 = (D_N)^2 = \left(\dfrac{2}{a+b}\hat{D}_N\right)^2. \tag{12}$$

2.3. Other Numerical Schemes

We mainly focus our attention on the numerical schemes of spatial discretization. The numerical method for time discretization is the classic Runge–Kutta method which is shown in Equation (4). The diffusive term "$\varepsilon \dfrac{\partial^2 f(L,t)}{\partial L^2}$" is no longer taken into account in the following two schemes. Thus, the right-hand term in Equation (4) becomes $R(t_n, f^n(L)) = h(t_n, f^n(L)) - \dfrac{\partial}{\partial L}(G^n(L)f^n(L))$.

2.3.1. Second Upwind Scheme

In the calculation of $\frac{\partial}{\partial L}(G^n(L)f^n(L))$, we use the following second-order upwind scheme [22]:

$$\left.\frac{\partial(G^n(L)f^n(L))}{\partial L}\right|_{L=L_i} = \begin{cases} \dfrac{3(Gf)_i^n - 4(Gf)_{i-1}^n + (Gf)_{i-2}^n}{2\Delta L} & G^n(L_i) \geq 0 \\ \dfrac{3(Gf)_{i+2}^n - 4(Gf)_{i+1}^n + (Gf)_i^n}{2\Delta L} & G^n(L_i) < 0 \end{cases}. \quad (13)$$

Here, f_i^n represents the value of f^n at node L_i, G_i^n represents the value of G^n at node L_i, and ΔL represents the distance between two adjacent nodes. During our simulation, the nodes are taken as uniformly distributed nodes, rather than Chebyshev points in the Chebyshev spectral collocation method.

2.3.2. HR-Van Scheme

In the calculation of $\frac{\partial}{\partial L}(G^n(L)f^n(L))$, the following HR-van scheme is used [1]:

$$\left.\frac{\partial(G^n(L)f^n(L))}{\partial L}\right|_{L=L_i} = \begin{cases} \frac{1}{\Delta L}[G_i^n\left(f_i^n + \frac{1}{2}\phi(r_i)(f_{i+1}^n - f_i^n)\right) \\ \quad -G_{i-1}^n\left(f_{i-1}^n + \frac{1}{2}\phi(r_{i-1})(f_i^n - f_{i-1}^n)\right)] & G^n(L_i) \geq 0 \\ \frac{1}{\Delta L}[G_{i+1}^n\left(f_{i+1}^n + \frac{1}{2}\phi(r_{i+1})(f_{i+2}^n - f_{i+1}^n)\right) \\ \quad -G_i^n\left(f_i^n + \frac{1}{2}\phi(r_i)(f_{i+1}^n - f_i^n)\right)] & G^n(L_i) < 0 \end{cases}, \quad (14)$$

where $r_i = \dfrac{f_i^n - f_{i-1}^n}{f_{i+1}^n - f_i^n}$, $\phi(r) = \dfrac{|r|+r}{1+|r|}$, with $\phi(r)$ being the flux limiter proposed by Van. Similarly, this scheme is implemented under uniformly distributed nodes rather than Chebyshev points.

3. Numerical Experiments

In order to compare the accuracy of each numerical method, we define the following errors:

$$L_1 - error = \sum_{k=0}^{N} |f_k^e - f_k^n|\Delta L_k / \sum_{k=0}^{N} |f_k^e|\Delta L_k$$
$$L_2 - error = \sqrt{\sum_{k=0}^{N}(f_k^e - f_k^n)^2 \Delta L_k} / \sum_{k=0}^{N} |f_k^e|\Delta L_k \quad (15)$$

Here, f_k^e is the exact solution at node L_k.

3.1. Size-Independent Growth in a Batch Process

The first example is the crystallization in a batch process with the crystal growth rate G independent of the crystal size L. Here, we do not consider the aggregation, nucleation, and breakage, and we set $h(L, t, f) = 0$. The crystal growth rate is set to $G = 1.0 \ \mu m/s$ and the initial CSD is given by with the following Gaussian distribution:

$$f(L, 0) = \frac{10^{10}}{\sqrt{2\pi}\sigma} e^{-\frac{(x-\mu)^2}{2\sigma^2}}, \quad (16)$$

with $\mu = 20 \ \mu m$, $\sigma = 3$. The other parameters in the calculation are set to $N = 200$, $dt = 0.001$, $\varepsilon = 0.005$, $t_{end} = 60 \ s$. The exact solution of this example is $f(L, t) = f(L - Gt, 0)$.

Figure 1 shows the comparison of exact solution with the results of three numerical scheme sat $t = 60$ s. In order to clearly show the performances with different schemes, Figure 1b shows the predicted crystal size distributions at the local crystal size. It can be seen that second-order upwind scheme is the worst, which causes both numerical dispersion and diffusion; both the HR-van scheme and the Chebyshev spectral collocation method cause numerical diffusion, but the Chebyshev spectral collocation method causes less numerical diffusion. Table 1 shows the L_1 and L_2 errors with different numerical schemes. The error table clearly shows that the Chebyshev spectral collocation method has the smallest error. Therefore, the Chebyshev spectral collocation method has the highest accuracy among these three methods.

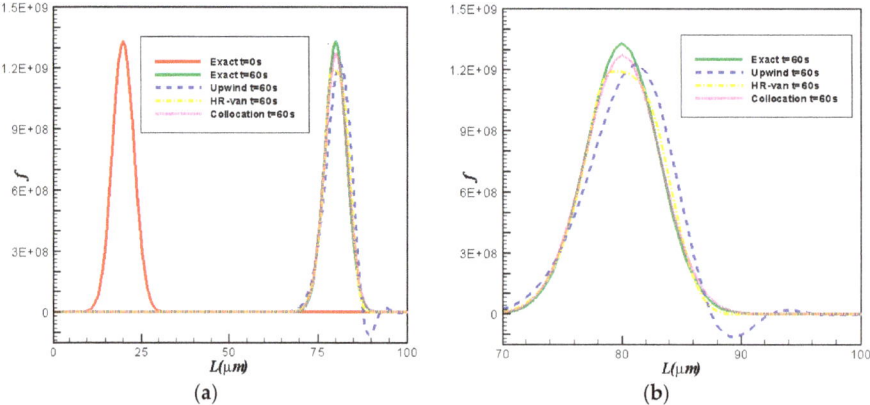

Figure 1. Comparison for final crystal size distribution (CSD) with analytical and numerical results: (a) in the whole crystal; (b) zoomed figure.

Table 1. L_1 and L_2 errors for different numerical schemes.

Method	Second Upwind	HR-Van	Collocation
L_1-error	0.24941	0.07598	0.04206
L_2-error	0.06334	0.02430	0.01265

We now discuss the effects of the diffusive term. Figure 2 shows the numerical results obtained using the Chebyshev spectral collocation method with different coefficients of diffusive term. As can been seen in Figure 2a, when no diffusive term is added ($\varepsilon = 0$), an oscillation happens near the right boundary at $t = 2$ s. Therefore, in this case, we cannot obtain stable numerical solutions. However, as shown in Figure 2b, when a diffusive term is added ($\varepsilon = 0.002$), a convergent solution is obtained. The CSD function shifts to the right with the velocity of G. Table 2 lists the L_1 and L_2 errors with different coefficients of the diffusive term. When $\varepsilon \leq 0.001$, a convergent numerical solution cannot be obtained. With the increase of ε ($\varepsilon \geq 0.002$), both L_1 and L_2 errors increase. This is consistent with our explanation for the diffusive term; the closer ε is to 0, the more accurate the result is. However, when ε approaches 0, the effect of the diffusive term becomes smaller which leads to an unstable numerical solution. The coefficient of the diffusive term (ε) with $G = 1$ is recommended as 2×10^{-3} to 10^{-2}.

In the size-independent growth case, PBE is a convection equation ($\frac{\partial f}{\partial t} + G\frac{\partial f}{\partial L} = 0$) and displaysa strong hyperbolic property. The diffusive term should be added in order to obtain a stable and convergent numerical solution. Figure 3 shows the relationship between the optimal ε(with minimum L_1 and L_2 errors) and the growth rate G. It is clear that the optimal ε has a linear relationship with the growth rate G. Therefore, the coefficient of diffusive term (ε) is recommended as $2G \times 10^{-3}$ to $G \times 10^{-2}$, where G is the crystal growth rate.

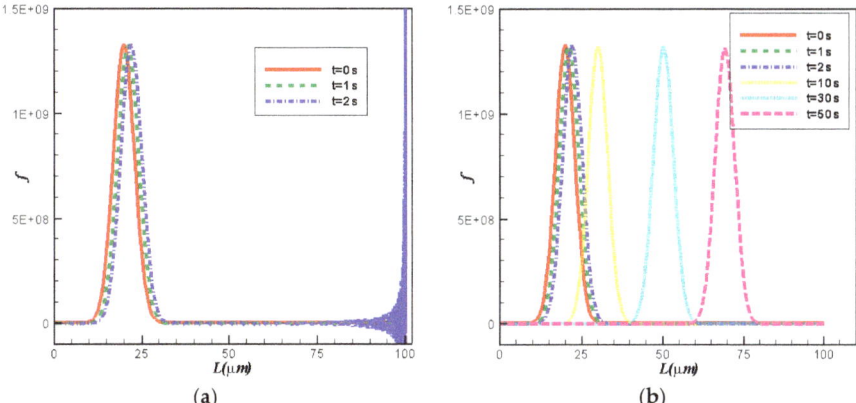

Figure 2. Comparison for numerical results with different coefficients of the diffusive term: (**a**) $\varepsilon = 0$; (**b**) $\varepsilon = 0.002$.

Table 2. L_1 and L_2 errors for different coefficients of the diffusive term.

ε	$\leq$	0.002	0.01	0.02
L_1-error	—	0.01286	0.06111	0.11548
L_2-error	—	0.00389	0.01828	0.03403

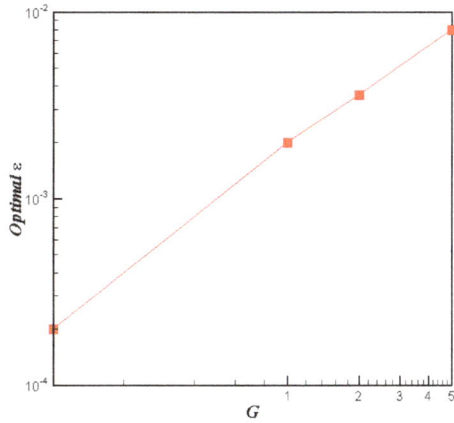

Figure 3. Relationship between optimal ε and G.

3.2. Size-Dependent Growth in a Batch Process

The second example involves crystallization in a batch process with the crystal growth rate G dependent on the crystal size L. Similar to the first example, aggregation, nucleation, and breakage are not taken into account, which means $h(L, t, f) = 0$. Crystal growth rate is supposed as the linear function of crystal size, namely $G(L, t) = G_0 L$, where G_0 is a constant. The initial CSD satisfies the following equation:

$$f(L, 0) = \frac{N_0}{\overline{L}} \exp\left(-\frac{L}{\overline{L}}\right), \qquad (17)$$

where $N_0, \bar{L}$ are constants. The analytical solution for this case is [1] as follows:

$$f(L,t) = \frac{N_0}{\bar{L}} \exp(-\frac{L}{\bar{L}}e^{-G_0 t}) \exp(-G_0 t). \tag{18}$$

Parameters in the simulation are taken as $\bar{L} = 0.01~\mu m^3, N_0 = 1, G_0 = 0.1~\mu m^3/s, \varepsilon = 0, N = 200, \Delta t = 0.00001~s$. The number of grids in space used for the second upwind method and HR-van method is 1000.

Figure 4 shows the comparison of analytical solutions with numerical solutions obtained by three numerical schemes at $t = 4~s$. In Figure 4a, there is no significant difference between the analytical solution and the numerical solutions. Figure 4b gives the comparison of the numerical results near the boundary $L = 0~\mu m$. From Figure 4b, it is obvious that Chebyshev spectral collocation method performs the best and its result is almost coincident with the analytical solution; however, the second-order upwind scheme and HR-van scheme have some deviations from the analytical solution near the boundary. Table 3 gives the comparison of L_1 and L_2 errors for different numerical schemes. From Table 3, it is clear that the Chebyshev spectral collocation method has the smallest error, whereas the HR-van scheme has a smaller error and the second-order upwind scheme has the biggest error. Thus, the Chebyshev spectral collocation method also has the highest accuracy among these three schemes.

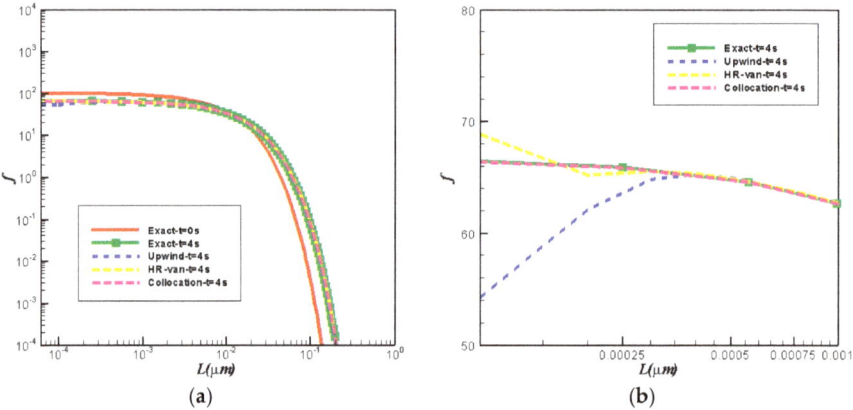

Figure 4. Comparison for final CSD with analytical and numerical results: (**a**) in the whole crystal; (**b**) zoomed figure.

Table 3. L_1 and L_2 errors for different numerical schemes.

Method	Second Upwind	HR-Van	Collocation
L_1-error	0.005036	0.004928	1.7992×10^{-7}
L_2-error	0.352802	0.328718	7.5825×10^{-7}

Figure 5 presents the evolution of CSD obtained using the Chebyshev spectral collocation method. As we can see in Figure 5, there is no oscillation during the computation. Therefore, the diffusive term is not necessary. This is not surprising. In this size-dependent growth case, the PBE is a convection–reaction equation ($\frac{\partial f}{\partial t} + G_0 L \frac{\partial f}{\partial L} + G_0 f = 0$). The reactive term weakens the hyperbolic property of PBE and makes it easy to solve. This is reflected in the decay term "$\exp(-G_0 t)$" of the analytical solution in Equation (18).

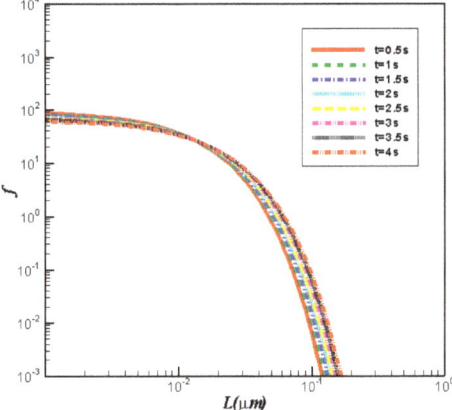

Figure 5. Evolution of CSD obtained using the Chebyshev spectral collocation method.

3.3. Nucleation and Size-Dependent Growth in a Continuous Process

The third example is the crystallization in a continuous process with the nucleation and crystal growth rate G dependent on the crystal size L. Here, the source item is assumed to $h(L, t, f) = B_0 \delta(L) - \dfrac{f(L)}{\tau}$, where B_0, τ are constants, and $\delta(L)$ is a Dirac δ function. The relationship between crystal growth rate and crystal size is $G(L) = G_0(1 + \gamma L)^z$, where G_0, γ, z are constants. The analytical solution for this case is [1] as follows:

$$\lim_{t \to \infty} f(L) = \frac{B_0}{G_0}(1+\gamma L)^{-z} \exp\left(\frac{1-(1+\gamma L)^{1-z}}{G_0 \tau \gamma (1-z)}\right). \tag{19}$$

In order to compare the prediction effects of each numerical method, Equation (19) is given as the initial CSD and the simulation is ended at $t = 10000$ s. In our simulation, parameters are set to $B_0 = 2 \times 10^{-10}/\mu m^3/s$, $\tau = 100$ s, $G_0 = 0.00168$ $\mu m/s$, $z = 0.3$, $\gamma = 1$, $\varepsilon = 0$, $N = 40$, $dt = 0.001$. The number of grids in space used for the second upwind method and HR-van method is 400.

Figure 6 shows the comparison of analytical solutions with numerical solutions of three numerical schemes. From Figure 6a,b, we can conclude that the result of the second-order upwind scheme is the worst since the predicted result has the biggest deviation from the analytical solution. Results of the HR-van method and the Chebyshev spectral collocation method are satisfactory. Figure 6b shows that the results obtained using the Chebyshev spectral collocation method are slightly better than those of the HR-van method. Table 4 shows the comparison of L_1 and L_2 errors for three different numerical schemes. From the error comparison, we know that the Chebyshev spectral collocation method is the best one, whereas the HR-van scheme is the better one and the second-order upwind scheme is the worst one. Thus, in this case, the Chebyshev spectral collocation method also shows the highest accuracy.

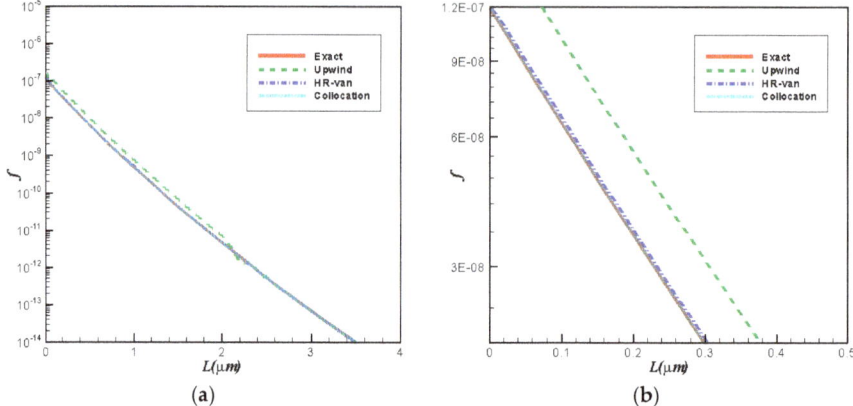

(a)   (b)

Figure 6. Comparison for final CSD with analytical and numerical results: (**a**) in the whole crystal; (**b**) zoomed figure.

Table 4. L_1 and L_2 errors for different numerical schemes.

Method	Second Upwind	HR-Van	Collocation
L_1-error	0.513976	0.0285996	3.559151×10^{-7}
L_2-error	0.856945	0.049325	4.422907×10^{-7}

Figure 7 displays the evolution of CSD obtained using the Chebyshev spectral collocation method. It can be seen that it does not result in an unstable solution in the case where no diffusive term is added. In this case, the PBE becomes $\frac{\partial f}{\partial t} + G\frac{\partial f}{\partial L} + f\frac{\partial G}{\partial L} = B_0\delta(L) - \frac{f(L)}{\tau}$. It is a convection–reaction equation. As shown in example 2, the reactive term weakens the hyperbolic property of PBE. Therefore, the diffusive term is not necessary.

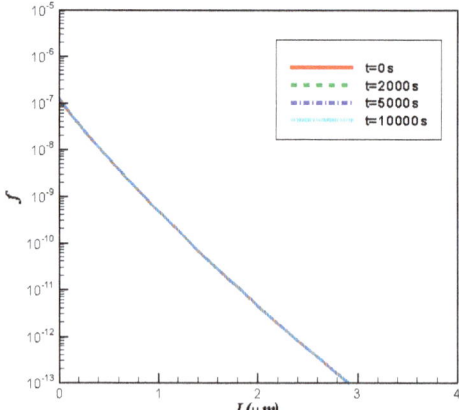

Figure 7. Evolution of CSD obtained using the Chebyshev spectral collocation method.

4. Conclusions

In this paper, the Chebyshev spectral collocation method was introduced to solve the one-dimensional PBE. The detailed implementation and the related techniques of the Chebyshev spectral collocation method are given. Three examples are presented, namely size-independent growth, size-dependent growth in a batch process, and with nucleation, and size-dependent growth in a continuous process. The numerical results showed that the Chebyshev spectral collocation method is effective and highly accurate in solving the PBE. However, in the case of size-independent growth, when PBE is a convection equation, it displays a highly hyperbolic property. A diffusive term should be added in order to obtain a stable solution.

For the more complex crystallization processes, e.g., aggregation and breakage, PBE becomes an integro-differential equation. Other methods, e.g., the cell average method and the fixed pivot method, can be combined with the spectral collocation method to obtain accurate CSD [15]. Moreover, realistic PBE problems involve multidimensional states. The spectral collocation method can be easily implemented for two-dimensional problems, enabling its use in multivariate systems. However, the spectral collocation method as a spectral method has its limitations in solving PBE, i.e., the computational zone should be simple and the distribution of CSD should be sufficiently smooth. Therefore, under such conditions, the Chebyshev spectral collocation method is recommended to solve PBE for which a high accuracy of CSD is needed.

Acknowledgments: Financial support was provided by the Natural Sciences Foundation of China (Nos. 11402078, 51375148, U1304521).

Conflicts of Interest: The authors declare no conflicts of interest.

References

1. Gunawan, R.; Fusman, I.; Braatz, R.D. High resolution algorithms for multidimensional population balance equations. *AICHE J.* **2004**, *50*, 2738–2749. [CrossRef]
2. Ramkrishna, D. *Population Balances: Theory and Application to Particulate Systems in Engineering*; Academic Press: New York, NY, USA, 2000.
3. Li, D.Y.; Li, Z.P.; Gao, Z.M. Quadrature-based moment methods for the population balance equation: An algorithm review. *Chin. J. Chem. Eng.* **2019**, in press. [CrossRef]
4. Qamar, S.; Angelov, I.; Elsner, M.P.; Ashfaq, A.; Seidel-Morgenstern, A.; Warnecke, G. Numerical approximations of a population balance model for coupled batch preferential crystallizers. *Appl. Numer. Math.* **2009**, *59*, 739–753. [CrossRef]
5. Qamar, S.; Ashfaq, A.; Angelov, I.A.; Elsner, M.P.; Warnecke, G.; Seidel-Morgenstern, A. Numercial solutions of population balance models in preferential crystallization. *Chem. Eng. Sci.* **2008**, *63*, 1342–11352. [CrossRef]
6. Solsvik, J.; Jakobsen, H.A. Evaluation of weighted residual methods for the solution of a population balance model describing bubbly flows: The least-squares, Galerkin, tau, and orthogonal collocation methods. *Ind. Eng. Chem. Res.* **2013**, *52*, 15988–16013. [CrossRef]
7. Sewerin, F.; Rigopoulos, S. An explicit adaptive grid approach for the numerical solution of the population balance equation. *Chem. Eng. Sci.* **2017**, *168*, 250–270. [CrossRef]
8. Shah, B.H.; Ramkrishna, D.; Borwanker, J.D. Simulation of particulate systems using the concept of the interval of quiescence. *AICHE J.* **1977**, *23*, 897–904. [CrossRef]
9. Gooch, J.R.P.; Hounslow, M.J. Monte Carlo simulation of size-enlargement mechanisms in crystallization. *AICHE J.* **1996**, *42*, 1864–1874. [CrossRef]
10. Kumar, S.; Ramkrishna, D. On the solution of population balance equations by discretization—I. a fixed pivot technique. *Chem. Eng. Sci.* **1997**, *52*, 1311–1332. [CrossRef]
11. Ma, D.L.; Tafti, D.K.; Braatz, R.D. High-resolution simulation of multidimensional crystal growth. *Ind. Eng. Chem. Res.* **2002**, *41*, 6217–6223. [CrossRef]
12. Majumder, A.; Kariwala, V.; Ansumali, S.; Rajendran, A. Lattice Boltzmann method for multi-dimensional population balance models in crystallization. *Chem. Eng. Sci.* **2012**, *70*, 121–134. [CrossRef]
13. Ruan, C.L.; Liang, K.F.; Chang, X.J.; Zhang, L. Weighted Essentially Nonoscillatory method for two-dimensional population balance equations in crystallization. *Math. Prob. Eng.* **2013**, *125128*, 1–11. [CrossRef]
14. Mantzaris, N.V.; Daoutidis, P.; Srienc, F. Numerical simulation of multi-variable cell population balance models. II. Spectral methods. *Compt& Chem. Eng.* **2001**, *25*, 1441–1462.

15. Buck, A.; Klaunick, G.; Kumar, J.; Peglow, M.; Tsotsas, E. Numerical simulation of particulate processes for control and estimation by spectral methods. *AICHE J.* **2012**, *58*, 2309–2319. [CrossRef]
16. Shen, J.; Tang, T.; Wang, L.L. *Spectral Methods: Algorithms, Analysis and Applications*; Springer: Berlin, Germany, 2011.
17. Zhang, X. *Matlab Efficient Solution of Differential Equation: SpectralMethod*; China Machine Press: Beijing, China, 2016.
18. Tian, X.Y.; Li, B.W.; Wu, Y.S.; Zhang, J.K. Chebyshev collocation spectral method simulation for the 2D boundary layer flow and heat transfer in variable viscosity MHD fluid over a stretching plate. *Int. J. Heat. Mass. Trans.* **2015**, *89*, 829–837. [CrossRef]
19. Chen, Y.Y.; Li, B.W.; Zhang, J.K. Spectral collocation method for natural convection in a square porous cavity with local thermal equilibrium and non-equilibrium models. *Int. J. Heat. Mass. Trans.* **2016**, *96*, 84–96. [CrossRef]
20. Fang, J.W.; Wu, B.Y.; Liu, W.J. An explicit spectral collocation method using nonpolynomial basis functions for the time-dependent Schrodinger equation. *Math. Meth. App. Sci.* **2019**, *42*, 186–203. [CrossRef]
21. Zhang, H.; Jiang, X.Y.; Wang, C.; Chen, S.Z. Crank-Nicolson Fourier spectral method for the space fractional nonlinear Schrodinger equation and its parameter estimation. *Int. J. Compt. Math.* **2019**, *96*, 238–263. [CrossRef]
22. Mark, H. *Introduction to Numerical Methods in Differential Equaitons*; Springer: Berlin, Germany, 2011.

© 2019 by the author. Licensee MDPI, Basel, Switzerland. This article is an open access article distributed under the terms and conditions of the Creative Commons Attribution (CC BY) license (http://creativecommons.org/licenses/by/4.0/).

Article

Shift Scheduling with the Goal Programming Method: A Case Study in the Glass Industry

Özlem Kaçmaz, Haci Mehmet Alakaş and Tamer Eren *

Department of Industrial Engineering, Kirikkale University, Kirikkale 71450, Turkey; ozlemkacmazz@hotmail.com (Ö.K.); hmalagas@gmail.com (H.M.A.)
* Correspondence: teren@kku.edu.tr; Tel.: +90-318-357-4242 (ext. 1050)

Received: 15 April 2019; Accepted: 17 June 2019; Published: 20 June 2019

Abstract: Nowadays, resource utilization and management are very important for businesses. They try to make a profit by providing high levels of efficiency from available sources. Their labor force is one of these sources. Therefore, scheduling based on personnel satisfaction has become an important problem in recent years. In this study, a case study was carried out in a glass factory in Ankara which has 7 department and 80 personnel. The aim of the study is to provide better service by distributing personnel to shifts in a fair and balanced manner. Assignment points are different for the departments where the personnel will work. Every personnel member is assigned to the department as best as possible. A goal programming method was used, and the results were better than those obtained using other methods.

Keywords: shift schedule; goal programming; labor; assignment; personnel

1. Introduction

Today, resource utilization and management is very important for businesses. They try to make a profit by providing high levels of efficiency from available sources. Their labor force is one of these sources. The human factor has different demands and expectations than other sources, and it increases the level of interest in this field. The shift scheduling problem, which is a sub-problem of personnel scheduling, is the most frequently studied problem from past to present. The shift scheduling problem, which is encountered in industries, enterprises, hospitals and many other places, is an important problem. To solve this problem, we aimed to provide better service by distributing personnel to shifts in a fair and balanced manner. In addition, the shift scheduling problem provides more income for institutions and organizations by improving employee wages, overtime and break times [1].

The problem becomes complicated if the solution to the shift scheduling problems is about increasing the satisfaction level of the employee and the enterprise. In terms of business, the seniority levels, knowledge and skills of the personnel are taken into consideration, and from the point of view of the personnel, expectations are met. Being one of the conditions that the parties want to provide for each other, it becomes difficult to solve the problem manually. Goal Programming is a mathematical programming method which aims to simultaneously provide multiple and conflicting goal constraints. As with other mathematical programming methods, the optimal solution is not the result. It works to minimize the deviation variables added to the goal constraints. So, the results are displayed as the closest to the targeted value or values.

In this study, 80 personnel and 7 departments (cutting (1), sanding (2), grinding (3), tempering (4), laminating (5), double glazing (6), shipment (7)) in a glass factory were used. In practice, the shifts of personnel working in the factory are scheduled. In the table consisting of two shifts, besides the provision of the number of personnel needed for the shift, the knowledge, skills and requests of the personnel were taken into consideration. The question was asked which personnel should be assigned to which shift for which department. Improvements were made according to the table prepared in the current situation.

In the second part of the study, shift scheduling is examined; in the third part, goal programming (GP) is discussed. The fourth part comprises a literature review, the fifth part a case study, and the last part presents the results.

2. Shift Scheduling

One situation in which staff scheduling problems are commonly encountered is shift scheduling. During the planning period, personnel perform certain activities such as work, rest, eating and taking tea breaks, week holidays and annual leave. These activities should occur within the framework of certain rules and laws. The equitable and balanced distribution of the personnel to the shifts is called shift scheduling. Businesses and institutions can make a profit by providing a good chart and a high yield.

Shift scheduling problems have been very widely discussed in the literature. The first integer mathematical model for shift scheduling was developed by George Dantzig in 1954, and the second by Elbridge Keith in 1979. In the academic literature, the model developed by Dantzig on shift scheduling and that developed by Keith in commercial studies have attracted the most attention [2].

In shift schedules, it is sometimes desirable to employ personnel mainly in the fields of business or departments where they are experts. Considering both the shift scheduling of these problems and the employment of the personnel in the fields where they are experts, it takes a long time to find an optimal solution. In addition, it is very difficult to deal with this problem manually, even by a skilled person. Developing a mathematical model for such problems is of benefit in every respect [3].

3. Goal Programming Method

Goal programming is a type of multi-purpose programming model. Model constraints are written by adding deviation variables to the targeted constraints, except for the indispensable constraints when creating the model. The aim is to minimize deviation variables in goal constraints. It is expected that multiple objectives will be provided at the same time in the models created using goal programming. At the same time, deviations of constraints can be minimized by converting these objectives into constraints and ranking them according to their importance [1].

It is not always possible to achieve every goal determined by the target programming method. Optimal results are selected from the among the most satisfactory. Targets are created for the selected solution. The priorities for the created target are determined. By performing these steps, the model is provided in general. Finally, the solution is determined.

The mathematical representation of the goal programming is as follows [4].

$$\text{Min } Z = \sum_{i=1}^{k}\left(d_i^+ + d_i^-\right), i = 1,\ldots,k \tag{1}$$

$$\sum_{j=1}^{n} a_{ij} X_j + d_i^- - d_i^+ = b_i, i = 1,\ldots,k, j = 1,\ldots,n \tag{2}$$

$$X_j, d_i^+, d_i^- \geq 0, i = 1,\ldots,k, j = 1,\ldots,n \tag{3}$$

Variables

X_j : j. decision variable, $j = 1,\ldots,n$
a_{ij} : coefficients of ith goal in variable j, $i = 1,\ldots,k$ $j = 1,\ldots,n$
b_i : desired goal value of the ith goal, $i = 1,\ldots,m$
d_i^+ : the deviation values in the positive directions from the ith goal, $i = 1,\ldots,m$
d_i^- : the deviation values in the negative directions from the ith goal, $i = 1,\ldots,m$

4. Literature Review

In the literature, there are many comprehensive studies on shift scheduling and staff/personnel scheduling. In [5], Gungor proposes an integer linear model for nursing scheduling in a hospital that is open 24 h a day, 7 days a week, where all nurses were staffed and worked for 40 h per week. The model consists of two stages. First, the minimum number of nurses that need to be fulfilled, and how many of them could be a student nurse is determined; then, a work and holiday schedule for a period of 2 weeks is designed. Bard et al. [6] modeled the tour scheduling problem in the United States Postal Service using integer programming. They added the restrictions set by a trade union agreement to the model. They presented scenarios that aimed to reduce the size of the workforce by producing solutions in 1 h. Wong and Chun [7] examined the nursing scheduling problem using a probability-based technique. As a result of their work, they brought solutions to the problem in a short time and presented appropriate charts. Azaiez and Sharif [8] developed a model for a computerized 0-1 Goal Programming method for nurse scheduling. This model is adapted to a hospital program in Saudi Arabia. The model prevents unnecessary overtime costs when considering hospital goals and nurse preferences. It is also implemented over a six-month period. Seckiner et al. [9] presented the hierarchical workforce scheduling problem with an integer programming model. When they compared the results between the current model and the solved model, they found that worker costs had been reduced. Personnel assigned to a single shift in the previous model can be assigned to alternate shifts by the proposed model. The model provides flexibility to the decision maker. Sungur, et al. [2], using the integer programming method, aimed to reduce labor costs and enabled employers to assign shifts in the most appropriate way. Castillo et al. [10] aimed to minimize labor costs by including service quality in their work at a call center, as well as achieving optimal staff scheduling. They put forward a multi-criteria paradigm. Olive [11] developed a mathematical model for the manufacturing industry to address personnel scheduling problems; to this end, he used mixed integer programming. The aim of this study is to meet the required number of work hours in different time periods. Topaloglu [12] discussed the problem of scheduling medical assistants in a hospital. Assistants are classified according to their seniority levels. The model is solved and scheduled for six months with goal programming. When the results are examined, better performance is shown compared to manual methods. Heimerl and Kolisch [13] discussed the labor planning problem in multiple projects using mixed integer linear programming. The aim of integer linear programming is to minimize labor costs. In terms of solution results, it has received better results than simple heuristic methods. Karaatli [14] studied the health sector and worked on the nursing scheduling problem in which work hours were 24/7. Fuzzy linear programming has benefited from genetic and heuristic algorithms; the results differed according to the methods. Koruca [15] developed a simulation-supported shift planning module in a small-scale enterprise manufacturing central heating boilers. In order to determine the situation, he undertook study and data collection. As a result of the study, four different shift plan scenarios suitable for possible crisis environments were presented. Atmaca et al. [16] determined the number of nurses who needed to be assigned to shifts in a hospital while increasing hospital efficiency, ensuring customer satisfaction and minimizing costs. Then, they compared the results obtained from current methods and the proposed model. Bag et al. [17] addressed the problem of nurse scheduling in a state hospital in Kirikkale using 0-1 goal programming and the Analytical Network Process (ANP) method. They solved the model with 0-1 goal programming and used ANP to determine the weight in goal programming. They used 5 goals in the study. They compared the solution results to the previous situation. Bektur and Hasgul [18] applied this technique to restaurant taking into account the staff, seniority levels and system staff. The objective function of the model is to minimize deviations from loose constraints according to seniority levels. The proposed model gave better results than current methods. Desert [19] examined the problem of labor recruitment using the goal programming method. A schedule for employees in a restaurant, seniority levels, preferences and shifts in accordance with the preferences of the day was made. Unal and Eren [20] addressed the problem of scheduling personnel in a government agency. They used the weighted goal programming method by considering the demands of the personnel, and solved the model with the GAMS 22.5 package program. As a result, the personnel were assigned according to their seniority levels, and most importantly, they were assigned shifts on their preferred days. More effective charts were prepared in a shorter time with the model. Ozcan et al. [4], in a large-scale hydroelectric power plant in Turkey, set a goal programming model by using real data. In the installed model, a 91% improvement was achieved in production stoppages caused by operator errors, taking into account the performance of employees and the requirements of the work. In this study, 3 goal constraints were determined. Ozder et al. [21] provided the best possible cleaning service to a 24-hour university hospital, proposing a monthly chart for 70 staff. They benefited from goal programming as a method. In [3], Varli devised the monthly working schedules of formen working in the bearing sector by considering different scenarios. Goal programming was used as the method. Attempts were made to take the wishes of the formen into account with the least possible level of deviation. The shifts, days and sections used in the scenarios are the same. The different formen numbers are special constraints and goal constraints. In [1], Varli and Eren studied the seniority levels of the workers in a factory using the Analytic Hierarchical Process (AHP) method. They then developed a model with goal programming to meet the number of employees needed for each shift and to make a distribution in a balanced and fair manner. Five goal constraints were used in the study. The same authors [22] aimed to ensure that the supervisory appointments of research assistants in the Faculty of Engineering of Kirikkale University were made in the most appropriate way during the final period. Goal programming was used as the method. Seventy four research assistants were assigned to 741 exams. They also [23] discussed the staff scheduling problem for nurses working in the internal medicine and endocrine departments of a hospital. For the monthly schedule to be created, hospital rules and special permission requests of the nurses were taken into consideration. Once again, the goal programming method was used. As a result of improvements and solutions, service quality is expected to increase. Bedir [24] aimed at reducing production downtime costs from personnel at a hydroelectric power plant by suggesting the use of the 0-1 Priority goal programming model considering personnel competencies. Competencies may be prioritized with the PROMETHEE method. The criteria affecting personnel competences were weighted using the AHP method. As a result of the solution, an 86% improvement was achieved for August 2017, i.e., when the plant was operating most intensively. The studies conducted in the literature have similarities with the present study. In our study, attention was given to the prioritization of personnel competencies. Gur and Eren [25] examined scheduling and planning problems using the goal programming method. As a result of the examination, they categorized the problems with a detailed analysis. Koc [26] aimed at minimizing all direct and indirect costs related to workforce scheduling. The method made use of integer programming. Koctepe [27] created a model for a meeting organization using the 0-1 integer programming method. In a model where personnel competencies are taken into consideration, planning was devised for 2 shifts and 80 personnel for 7 days. As a result of that study, it was determined that personnel satisfaction had increased. Tapkan [28] discussed the task scheduling problem of the Kayseri rail transportation system. Again, the multipurpose 0-1 mixed integer model was used. In the objective function of the mathematical model, the largest difference between the number of staff, the weekly statutory working time of the weekly working hours of the staff, the sum of the overrun periods and the average rest period and the rest period were taken into account. Ozder et al. [29] changed the shifts of personnel using ANP and goal programming methods in a natural gas combined cycle power plant. In order to incorporate the personnel skills into the model, they calculated the seniority levels with the ANP method. They identified 4 levels of seniority among 80 staff members for 3 shifts. The aforementioned studies are presented in Table 1.

Table 1. Literature review.

Author	Type	Methods
Varlı and Eren [1]	Shift scheduling	AHP, Goal Programming
Sungur [2]	Shift scheduling	Fuzzy Integer Programming
Varlı [3]	Shift scheduling	AHP, Goal Programming
Ozcan et al. [4]	Shift scheduling	Goal Programming
Güngör [5]	Personnel Scheduling	Integer Programming
Bard et al. [6]	Personnel Scheduling	Integer programming
Wong and Chun [7]	Personnel Scheduling	Constraint programming and Heuristic Method
Azaiez and Al Sharif [8]	Personnel Scheduling	0-1 Goal Programming
Seçkiner et al. [9]	Shift scheduling	Integer programming
Castillo et al. [10]	Shift scheduling	labor scheduling paradigm
Olive [11]	Shift scheduling	Mathematical model
Topaloglu [12]	Shift scheduling	Multiple objective programming
Heimerl and Kolisch [13]	Project scheduling	Integer programming
Karaatlı [14]	Personnel Scheduling	Linear Programming, Genetic and Heuristic Algorithm
Koruca [15]	Shift scheduling	simulation
Atmaca et al. [16]	Personnel Scheduling	0-1 linear goal programming
Bag et al. [17]	Personnel Scheduling	0-1 goal programming, ANP method
Bektur and Hasgul [18]	Shift scheduling	Goal Programming
Çöl [19]	Shift scheduling	Goal Programming
Unal and Eren [20]	Shift scheduling	Goal Programming, Multiple-Objective Decision Making
Ozder et al. [21]	Personnel Scheduling	Goal Programming
Varlı et al. [22]	Personnel Scheduling	Goal Programming
Varlı et al. [23]	Personnel Scheduling	Goal Programming
Bedir [24]	Shift scheduling	Goal Programming, AHP, PROMETHEE
Gur and Eren [25]	Literature Review	Research
Koc [26]	Shift scheduling	Integer Programming
Koctepe et al. [27]	Fair Scheduling	Integer Programming
Tapkan et al. [28]	Personnel Scheduling	0-1 mixed integer model
Ozder et al. [29]	Shift scheduling	Goal Programming, ANP

5. A Case Study in a Glass Factory in Ankara Province

The present study was carried out in a glass factory operating in the facade sector in Ankara in order to ensure the optimal appointment of personnel to shifts and departments. The factory workday comprises two shifts, i.e., 08:00-18:00 and 22:00-08:00. Personnel change shifts at one week intervals; the factory is closed on Sundays. Seven sections and 80 personnel, in which production goes on continuously, are discussed. When evaluating the existing system, it was assumed that the department and staff were working at full capacity. The relevant sections are cutting (1), sanding (2), grinding (3), tempering (4), laminating (5), double glazing (6) and shipment (7).

Orders from customers are numbered according to the type of product, in other words, according to the last process before the merchandise is made ready for shipment. Materials progress through the work order between the processes. The product range includes flat glass, colored glass, solar and temperature controlled glass, tempered glass, laminated glass and bulletproof glass.

5.1. Product Type

❖ <u>Flat Glass</u>: Flat glass has high light transmittance due to its transparency.
❖ <u>Colored Glass</u>: Colored glass is obtained by adding colorants to the glass paste; available in green, smoked, bronze and blue.
❖ <u>Solar and Heat Controlled Glass</u>: This is a type of glass with different aesthetics and designs that can save energy.
❖ <u>Tempered Glass</u>: A type of glass whose durability and resistance to thermal stresses are 5 times higher than those of flat glass. Areas of application are generally glass railings and doors, walk-in showers, intermediate compartments, glass furniture, refrigerator and oven windows, and side and rear windows of automobiles.

❖ Laminated Glass: Two or more glass plates are produced by combining special binder polyvinyl butyral (PVB) layers under heat and pressure. This process minimizes the risk of glass breakage by keeping the pieces in place in such an event. It contributes to sound insulation.
❖ Bullet Proof Glass: Bulletproof glass is aimed at preventing crime and facilitating the capture of the criminal after the action. Areas of use are banks, police stations, museums, military buildings and other official organizations, psychiatric wards, jewelers and so on. This category comprises polyvinyl butyral (PVB) or polycarbonate interlayer laminated glass.

5.2. Production Rotation

The order number is determined according to the process after which the glass will be ready for dispatch. Routes are created according to the following order numbers. The routes to be determined are cutting, grinding, tempering, laminating and double glazing. Glass is usually prepared by following these routes, except for cases of special orders. The following routes are shown in a flow chart. Routes of the products are shown in the flow charts given in Figure 1 for cutting, Figure 2 for grinding, Figure 3 for temper, Figure 4 for laminate and Figure 5 for double glazing glass.

Cutting Section Route; Cutting-Shipment

Figure 1. Cutting flow chart.

Grinding Section Route; Cutting- Grinding-Shipment

Figure 2. Grinding flow chart.

Temper Section Route; Cutting-Sanding Or Grinding-Tempering-Shipment

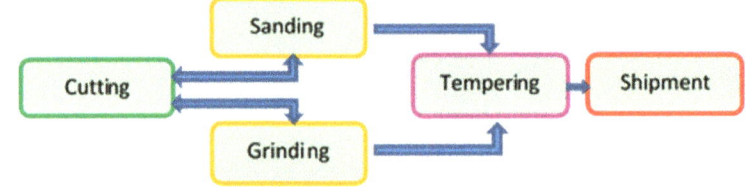

Figure 3. Tempering flow chart.

Laminating Section Route; Cutting-Sanding Or Grinding-Tempering-Laminated-Shipment

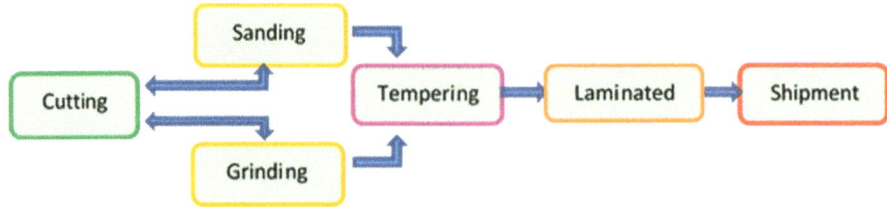

Figure 4. Laminating flow chart.

Double Glazing Section Route; Cutting-Sanding Or Grinding-Tempering-Double Glazing-Shipment

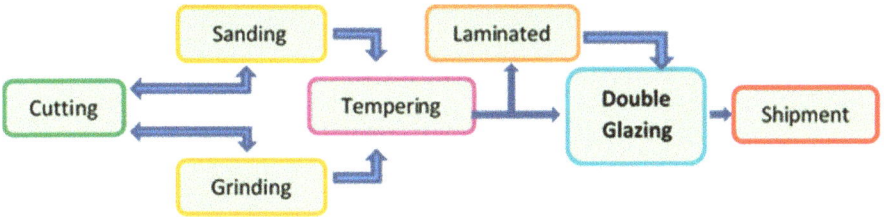

Figure 5. Double glazing flow chart.

5.3. List of Personnel

The factory has a total of 80 personnel. In practice, each personnel member receives points for each department. These scores are determined according to the opinions of experts. There scale is 1 to 3 points, according to the sections to be assigned. In the determined section, we tried to restricted to system to a maximum of 5 points. The model uses the minimum objective function; 1 point indicates that the competence level is greater than 2, while 3 points indicates that an employee is not competent in that section. The points obtained according to the personnel list and the sections are given in Table A1 in Appendix A section.

5.4. Mathematical Model

5.4.1. Parameters

n: number of personnel working in the factory, $n = 80$
m: number of days, $m = 30$
s: Number of sections in the factory, $s = 7$
t: Number of shifts, $t = 2$
i: Personnel index, $i = 1,2,\ldots,n$
j: Day index, $j = 1,2,\ldots,m$
k: Section index, $k = 1,2,\ldots,s$
l: Shift index, $l = 1,2,\ldots,t$

5.4.2. Decision variables

$$X_{ijkl} = \begin{cases} 1, \text{if shift, chapter and day is chosen for personnel} \\ 0, \text{otherwise} \end{cases}, i=1,2,\ldots,n, j=1,2,\ldots,m, k=1,2,\ldots,s, l=1,2,\ldots,t \quad (4)$$

$$h_{ij} = \begin{cases} 1, \text{if vacation for personnel} \\ 0, \text{otherwise} \end{cases}, i=1,2,\ldots,n, j=1,2,\ldots,m \quad (5)$$

5.4.3. Constraints

1-To meet the daily personnel needs of the departments:
Number of personnel needed for each shift in the cutting(1) section.

$$\sum_{i=1}^{n}(X_{ij1l}) = 3, j = 1,2,\ldots,m, l = 1,2 \quad (6)$$

Number of personnel needed for each shift in the sanding(2) section.

$$\sum_{i=1}^{n}(X_{ij2l}) = 4, j = 1,2,\ldots,m, l = 1,2 \quad (7)$$

Number of personnel needed for each shift in the grinding(3) section.

$$\sum_{i=1}^{n}(X_{ij3l}) = 4, j = 1, 2, \ldots, m, l = 1, 2 \tag{8}$$

Number of personnel needed for each shift in the tempering (4) section

$$\sum_{i=1}^{n}(X_{ij4l}) = 4, j = 1, 2, \ldots, m, l = 1, 2 \tag{9}$$

Number of personnel needed for each shift in the laminating (5) section

$$\sum_{i=1}^{n}(X_{ij5l}) = 6, j = 1, 2, \ldots, m, l = 1, 2 \tag{10}$$

Number of personnel needed for each shift in the double glazing (6) section

$$\sum_{i=1}^{n}(X_{ij6l}) = 8, j = 1, 2, \ldots, m, l = 1, 2 \tag{11}$$

Number of personnel needed for each shift in the shipment (7) section

$$\sum_{i=1}^{n}(X_{ij7l}) = 5, j = 1, 2, \ldots, m, l = 1, 2 \tag{12}$$

2- Only one shift per personnel per day:

$$\sum_{l=1}^{t}\sum_{k=1}^{s}(X_{ijkl}) \leq 1, i = 1, 2, \ldots, n, j = 1, 2, \ldots, m \tag{13}$$

3- personnel not working on the day of leave:

$$\sum_{l=1}^{t}\sum_{k=1}^{s}(X_{ijkl}) \leq (1 - h_{ij}), i = 1, 2, \ldots, n, j = 1, 2, \ldots, m \tag{14}$$

4- Each personnel member has a minimum of 1 and a maximum of 2 days a week.:

$$h_{ij} + h_{i(j+1)} + h_{i(j+2)} + h_{i(j+3)} + h_{i(j+4)} + h_{i(j+5)} + h_{i(j+6)} \leq 2, i = 1, 2, \ldots, n, j = 1, 2, \ldots, m-6 \tag{15}$$

$$h_{ij} + h_{i(j+1)} + h_{i(j+2)} + h_{i(j+3)} + h_{i(j+4)} + h_{i(j+5)} + h_{i(j+6)} \geq 1, i = 1, 2, \ldots, n, j = 1, 2, \ldots, m-6 \tag{16}$$

5-Upper limit for each personnel to work on 1 and 2 shifts:

$$\sum_{j=1}^{m}\sum_{k=1}^{s}(X_{ijk1}) \leq 12, i = 1, 2, \ldots, n \tag{17}$$

$$\sum_{j=1}^{m}\sum_{k=1}^{s}(X_{ijk2}) \leq 12, i = 1, 2, \ldots, n \tag{18}$$

6- Lower limit restrictions for each personnel on 1 and 2 shifts:

$$\sum_{j=1}^{m}\sum_{k=1}^{s}(X_{ijk1}) \geq 10, i = 1, 2, \ldots, n \tag{19}$$

$$\sum_{j=1}^{m}\sum_{k=1}^{s}(X_{ijk2}) \geq 10, i = 1, 2, \ldots, n \tag{20}$$

7- If an employee were assigned to the night shift on a given day, the next day's shift in the morning shift would be limited:

$$\sum_{k=1}^{s}(X_{ijk2}) + (X_{i(j+1)k1}) \leq 1, i = 1, 2, \ldots, n, j = 1, 2, \ldots, m-1 \tag{21}$$

5.4.4. Goal Constraints

Goal 1: Goal constraint where personnel are asked to minimize the assignment as day of leave-workday-leave when being assigned shifts:

$$h_{ij} + \sum_{k=1}^{s}\sum_{l=1}^{t}(X_{i(j+1)kl}) + h_{i(j+2)} + d_{1ij}^- - d_{1ij}^+ = 2, i = 1, 2, \ldots, n, j = 1, 2, \ldots, m-2 \quad (22)$$

Goal 2: Goal constraint where personnel are asked to minimize the assignment of working day-tracking-working day when being assigned to shifts:

$$\sum_{k=1}^{s}\sum_{l=1}^{t}(X_{ijkl}) + h_{i(j+1)} + \sum_{k=1}^{s}\sum_{l=1}^{t}(X_{i(j+1)kl}) + d_{2ij}^- - d_{2ij}^+ = 2, i = 1, 2, \ldots, n, j = 1, 2, \ldots, m-2 \quad (23)$$

Goal 3: Goal constraint on which the total number of vacancies for which each personnel is assigned is intended to be as equal as possible:

$$\sum_{j=1}^{m}\sum_{k=1}^{s}\sum_{l=1}^{t}(X_{ijkl}) + d_{3i}^- - d_{3i}^+ = 22, i = 1, 2, \ldots, n \quad (24)$$

Goal 4: Personnel assigned to the departments in each shift will provide the required sum of points as a qualification:

$$\sum_{i=1}^{n}(X_{ij1l}) * (1) + d_{4jl}^- - d_{4jl}^+ = 3, j = 1, 2, \ldots, m, l = 1, 2 \quad (25)$$

$$\sum_{i=1}^{n}(X_{ij2l}) * (2) + d_{5jl}^- - d_{5jl}^+ = 4, j = 1, 2, \ldots, m, l = 1, 2 \quad (26)$$

$$\sum_{i=1}^{n}(X_{ij3l}) * (3) + d_{6jl}^- - d_{6jl}^+ = 4, j = 1, 2, \ldots, m, l = 1, 2 \quad (27)$$

$$\sum_{i=1}^{n}(X_{ij4l}) * (4) + d_{7jl}^- - d_{7jl}^+ = 4, j = 1, 2, \ldots, m, l = 1, 2 \quad (28)$$

$$\sum_{i=1}^{n}(X_{ij5l}) * (5) + d_{8jl}^- - d_{8jl}^+ = 6, j = 1, 2, \ldots, m, l = 1, 2 \quad (29)$$

$$\sum_{i=1}^{n}(X_{ij6l}) * (6) + d_{9jl}^- - d_{9jl}^+ = 8, j = 1, 2, \ldots, m, l = 1, 2 \quad (30)$$

$$\sum_{i=1}^{n}(X_{ij7l}) * (7) + d_{10jl}^- - d_{10jl}^+ = 5, j = 1, 2, \ldots, m, l = 1, 2 \quad (31)$$

5.4.5. Objective Function

$$\text{Min} Z = \sum_{i=1}^{n}\sum_{j=1}^{m}(d_{1ij}^- + d_{1ij}^+) + (d_{2ij}^- + d_{2ij}^+) + (d_{3i}^- + d_{3i}^+) + \sum_{j=1}^{m}\sum_{l=1}^{t} d_{4jl}^+ + d_{5jl}^- + d_{6jl}^+ + d_{7jl}^+ + d_{8jl}^+ + d_{9jl}^+ + d_{10jl}^+ \quad (32)$$

5.5. Model Solution

First of all, general constraints and goal constraints are determined for the mathematical model. The current system was examined and the number of personnel needed by the departments was changed. Personnel are also allowed to leave on Sundays. Deviations in goal constraints were minimized and the objective function was created. The solutions used in the solution of the mathematical models were obtained using ILOG version 12.6.2. As a result of the solution, 30-day work schedules for 80 personnel were created. The suggested data is given in Table A2 in Appendix B. The solution produced using the proposed model was compared to that produced using current methods. The total competence score of the personnel working in the departments was divided by the number of people required in that department, and the shift labor force was created. The current system is given in Table 2 and the proposed system is given in Table 3.

Table 2. Average workforce in the current system.

Section	Cutting		Sanding		Grinding		Tempering		Laminated		Double Glazing		Shipment	
Shift \ Day	1	2	1	2	1	2	1	2	1	2	1	2	1	2
1	2.75	3	2.5	2.75	2.75	2.75	3.2	2.6	3	3.14	2.9	2.9	2.67	3.17
2	2.75	3	2.5	2.75	2.75	2.75	3.2	2.6	3	3.14	2.9	2.9	2.67	3.17
3	2.75	3	2.5	2.75	2.75	2.75	3.2	2.6	3	3.14	2.9	2.9	2.67	3.17
4	2.75	3	2.5	2.75	2.75	2.75	3.2	2.6	3	3.14	2.9	2.9	2.67	3.17
5	2.75	3	2.5	2.75	2.75	2.75	3.2	2.6	3	3.14	2.9	2.9	2.67	3.17
6	2.75	3	2.5	2.75	2.75	2.75	3.2	2.6	3	3.14	2.9	2.9	2.67	3.17
7	Staff on Leave													
8	3	2.75	2.75	2.5	2.75	2.75	2.6	3.2	3.14	3	2.9	2.9	3.17	2.67
9	3	2.75	2.75	2.5	2.75	2.75	2.6	3.2	3.14	3	2.9	2.9	3.17	2.67
10	3	2.75	2.75	2.5	2.75	2.75	2.6	3.2	3.14	3	2.9	2.9	3.17	2.67
11	3	2.75	2.75	2.5	2.75	2.75	2.6	3.2	3.14	3	2.9	2.9	3.17	2.67
12	3	2.75	2.75	2.5	2.75	2.75	2.6	3.2	3.14	3	2.9	2.9	3.17	2.67
13	3	2.75	2.75	2.5	2.75	2.75	2.6	3.2	3.14	3	2.9	2.9	3.17	2.67
14	Staff on Leave													
15	2.75	3	2.5	2.75	2.75	2.75	3.2	2.6	3	3.14	2.9	2.9	2.67	3.17
16	2.75	3	2.5	2.75	2.75	2.75	3.2	2.6	3	3.14	2.9	2.9	2.67	3.17
17	2.75	3	2.5	2.75	2.75	2.75	3.2	2.6	3	3.14	2.9	2.9	2.67	3.17
18	2.75	3	2.5	2.75	2.75	2.75	3.2	2.6	3	3.14	2.9	2.9	2.67	3.17
19	2.75	3	2.5	2.75	2.75	2.75	3.2	2.6	3	3.14	2.9	2.9	2.67	3.17
20	2.75	3	2.5	2.75	2.75	2.75	3.2	2.6	3	3.14	2.9	2.9	2.67	3.17
21	Staff on Leave													
22	3	2.75	2.75	2.5	2.75	2.75	2.6	3.2	3.14	3	2.9	2.9	3.17	2.67
23	3	2.75	2.75	2.5	2.75	2.75	2.6	3.2	3.14	3	2.9	2.9	3.17	2.67
24	3	2.75	2.75	2.5	2.75	2.75	2.6	3.2	3.14	3	2.9	2.9	3.17	2.67
25	3	2.75	2.75	2.5	2.75	2.75	2.6	3.2	3.14	3	2.9	2.9	3.17	2.67
26	3	2.75	2.75	2.5	2.75	2.75	2.6	3.2	3.14	3	2.9	2.9	3.17	2.67
27	3	2.75	2.75	2.5	2.75	2.75	2.6	3.2	3.14	3	2.9	2.9	3.17	2.67
28	Staff on Leave													
29	2.75	3	2.5	2.75	2.75	2.75	3.2	2.6	3	3.14	2.9	2.9	2.67	3.17
30	2.75	3	2.5	2.75	2.75	2.75	3.2	2.6	3	3.14	2.9	2.9	2.67	3.17

In the current system, it is assumed that the factory was operating at full capacity. Staff are only allowed a day off on Sundays. Persons are generally requested to work in the department in which they are employed. A task assignment is made to meet needs rather than special talents; that is why the average labor force is constant for each department, every week and for every shift.

For a sample average workforce calculation from the table, the 8th day shift of the sanding section will be considered; on that day, 2, 18, 34, 75 staff members were assigned. The total score of the assigned personnel for the sanding section was 2 + 2 + 3 + 3 = 10. The score obtained was divided by the number of staff required by the department, and the average workforce was found. This is equal to 10/4 = 2.5.

Table 3. Average workforce in the proposed model.

Section	Cutting		Sanding		Grinding		Tempering		Laminated		Double Glazing		Shipment	
Shift \ Day	1	2	1	2	1	2	1	2	1	2	1	2	1	2
1	1.67	2.33	1	1.5	2	1	2	1	1.83	1.67	1.38	1.75	1	1
2	1	3	1.5	1.75	1.5	1.75	2	1	2.67	1.83	1.5	2	1.2	2.8
3	1	3.67	1.5	1.25	1.25	1.5	2	1.5	1.67	1.67	1.25	1.25	1	1
4	1	2.33	1.5	1.25	1	3.75	1.5	1.5	1.33	2.33	1.5	2.5	1	1
5	2.33	1	1.25	1.25	2.25	1.25	1	2	1.17	1.67	1.5	1.5	1.4	1
6	2.33	2.33	1	1	1.75	1.25	1.25	1	1.17	1.67	2	1.38	1.8	2
7	3.67	1.67	1.25	1	1.5	2	2	1.5	1.33	2.17	2	2.25	1.6	1.4
8	1	2.33	1	1	1.25	1.25	2	1.5	1.67	1.33	2	1	1	1.2
9	1	2.33	1.5	1.25	1.5	1	2	1	1.67	1.67	1.5	1.88	1	1.8
10	1	2.33	1	1.25	1.5	1	2	2.25	1.83	2	1.25	2	1.2	1
11	2.33	1.67	1.5	1	2.5	2	2	1.25	1.33	2.67	1.75	1.13	1.4	1
12	3.67	1	1.25	1	1.5	1.5	2	1.75	2	1	1.75	2.13	1	1.6
13	3	2.33	1.25	1	1.5	1.5	1	1	1.17	1.33	2.75	1.5	2	1.4
14	3.67	1.67	1.25	1	1	1	1.5	1	1.67	1.67	1	1.25	1	1.6
15	2.33	1	1.25	1	1.5	1	2.5	1	1	1.67	1.5	1.63	1	1.4
16	3.67	1	1.75	1.25	1	3	1	2	1.33	1.5	2	2.13	1.2	1.8
17	1.67	2.33	1.25	2.25	1.5	1	2.25	1.5	1.33	1	1.5	1.38	1	1.2
18	1.67	1.33	1	1.25	1.5	2	1.25	1	1.5	1	1.75	1	1.4	1
19	1.67	2	1.25	1.25	2	2	1	2.25	1.33	1.33	1.63	2.13	1.4	1
20	2.33	2.33	1.5	1.25	2.25	1.25	2	2	1.33	1.33	1.5	1.5	1.2	1.4
21	3	2.33	1.5	1	1.5	1	1.25	2	2	1.83	1.5	1.25	1.2	1.6
22	3	2.33	1.25	2.25	1.25	3	2	2	1.33	1.33	1.5	1.25	1.2	2
23	1	2.33	1	1	2	2.25	1	2.5	1.33	1	2	1.5	1.2	1
24	1.67	2.33	1	1.25	1.25	2.5	1	1.5	1.83	1.67	1.25	1.75	1.8	1.8
25	2.33	1	1	1	1	1.25	1	1	1	1.33	1	1.25	1	1
26	1	1.67	1	1.25	1	1.5	1	1	1.33	1.33	2.38	1	1	1
27	1	2.33	1.25	1	1	1.25	3	2.5	1.5	1.83	2	2	1.4	1.4
28	3	1	1.5	1	1	1.5	1.5	2	1	1.33	2	2	1.8	1.8
29	4.33	2.33	1.25	1.25	1.5	1.25	1.5	2.5	1.33	1	1.25	1.75	1.8	1.8
30	1.67	1.67	1.75	1.25	1	1	1	2	1.67	2	1.5	1.5	1	1

With the proposed model, specific qualification scores have been defined for each department. Personnel can be assigned to different departments to provide a qualification score. The flexibility of work on different parts by the person provided flexibility for the day of leave. Thanks to our model, the shortage of authorized personnel was addressed. Thus, production can continue without stopages. The average workforce changes daily for each shift.

6. Conclusions

In the factory, we aimed to determine morning and evening shifts for 7 personnel and 80 personnel. Firstly, the number of personnel needed by each departments is emphasized. The number of personnel changed by taking into consideration that the right employee needs to be assigned to the appropriate department. In the current system, 4 staff are required for cutting, sanding, and grinding, 5 are needed for tempering, 7 for laminating, 10 for heaters and 6 for shipments. In the proposed system, these numbers are 3 for cutting, sanding, grinding, 4 for tempering, 6 for laminating, 8 for heaters and 5 for shipments. On Sunday, the personnel permits were distributed. Thus, the factory, which is in need of production, was allowed to operate on Sundays. In terms of the average labor force, only 5 of the 420 shifts did not achieve the desired result within a monthly planning period. In other shifts, the result was quite successful compared to the current situation. Every personnel member was assigned to only one shift during the day and no appointment was made on that employee's day of leave. Personnel were rated between 1 and 3 according to the departments to be assigned. One point indicates that the employee is more than competent for that part, 2 points indicates that he/she is moderately competent, and 3 points indicates insufficient competence. The assignment was restricted by giving 5 points for cases where personnel appointments were not requested. The upper and lower limit numbers can be determined and the distribution is equal for the morning and evening shifts. The personnel who worked a night shift were prevented from being assigned to the day shift the next day, as this would cause require more than 24 h of work withour a rest. In addition to this, it was desireable that the total number of shifts assigned to the personnel during the one-month planning period be the same, and that these appointments should be sequential around the day of leave. Finally, in order to achieve a certain level of points among the personnel assigned to the departments, a specific score for each department was given. The targets were gathered in one place and the objective function was established. Deviations from the target values determined by the goal programming method were very small. In this way, the most suitable shifts and departments could be assigned on the appropriate days by taking into consideration the talents of the personnel. The aim of the goal programming method is to perform multiple goals simultaneously.

In the existing system used in the factory, personnel are divided into two separate teams; each shift team is divided into 7 sections. The teams change at one week intervals. Each team should comprise only from competent people. Generally, a sufficient number of qualified persons are assigned to each section and the remaining personnel are used to complete the task; however, in this way, full efficiency cannot be achieved. The competence of the departments was improved with the proposed new model. Personnel were used more effectively being better assigned, and the satisfaction levels of the personnel increased. As a result, it was observed that the new system offers better results.

Author Contributions: Ö.K. conceptualized the study, prepared dataset, conducted analyses, contributed to writing the manuscript, and provided modeling processes. H.M.A. supported the solution process of the mathematical model and was a contributor to writing the manuscript. T.E. provided overall guidance and expertise in conducting the analysis and was a contributor to writing the manuscript. All authors read and approved the final manuscript.

Funding: This research received no external funding.

Conflicts of Interest: The authors declare no conflict of interest.

Appendix A

Table A1. The points of the personnel according to the sections.

	Cutting(1)	Sanding(2)	Grinding(3)	Tempering(4)	Laminated(5)	Double Glazing(6)	Shipment(7)
P1	1	1	1	5	5	5	2
P2	5	2	2	1	5	5	3
P3	5	1	1	3	5	3	5
P4	5	1	2	2	5	5	3
P5	3	3	5	1	5	5	5
P6	2	5	5	5	1	5	3
P7	5	1	1	2	3	5	5
P8	5	3	3	2	5	1	3
P9	3	2	2	5	1	5	5
P10	5	2	3	5	1	5	1

Table A1. *Cont.*

	Cutting(1)	Sanding(2)	Grinding(3)	Tempering(4)	Laminated(5)	Double Glazing(6)	Shipment(7)
P11	5	1	2	3	3	5	5
P12	5	1	1	5	5	5	3
P13	1	1	2	2	5	5	3
P14	5	1	2	1	5	5	5
P15	5	1	2	5	1	5	3
P16	5	2	3	1	5	5	1
P17	5	2	3	5	1	5	2
P18	3	2	5	2	5	1	5
P19	5	2	5	2	5	5	1
P20	5	1	2	5	5	2	1
P21	5	2	1	3	5	5	3
P22	1	1	1	5	3	5	2
P23	5	1	3	1	5	5	3
P24	5	2	3	2	5	1	5
P25	5	2	5	1	5	3	1
P26	3	1	2	5	3	5	3
P27	3	2	1	5	3	5	5
P28	5	5	5	3	1	5	1
P29	5	2	5	3	2	1	5
P30	5	2	1	3	5	5	2
P31	3	3	2	5	1	5	5
P32	1	5	5	3	2	3	1
P33	2	1	5	5	5	1	2
P34	5	3	1	5	5	3	5
P35	5	1	2	5	2	5	1
P36	3	5	5	1	3	1	1
P37	5	1	2	1	5	5	2
P38	1	2	3	3	5	5	3
P39	3	1	2	5	3	5	5
P40	5	2	1	5	5	5	1
P41	5	2	3	1	5	1	5
P42	5	1	1	1	3	5	5
P43	3	1	2	5	5	3	2
P44	3	1	1	5	5	3	5
P45	5	2	1	5	5	1	3
P46	5	2	1	3	3	5	5
P47	3	2	3	1	5	1	5
P48	5	2	3	2	1	5	5
P49	5	2	3	1	3	5	5
P50	5	2	1	5	5	1	2
P51	3	2	3	1	5	5	5
P52	5	1	2	3	5	3	5
P53	5	2	2	5	2	1	5
P54	5	1	2	5	1	5	5
P55	5	5	5	2	2	5	1
P56	3	2	2	1	5	5	3
P57	5	1	1	3	3	5	5
P58	3	1	1	5	3	5	2
P59	1	3	2	5	2	5	3
P60	3	5	5	2	5	5	1
P61	1	1	5	3	3	5	3
P62	5	2	5	5	1	3	5
P63	5	2	2	1	3	5	5
P64	5	2	3	5	3	1	1
P65	5	5	5	1	5	3	1
P66	3	1	1	5	5	5	3
P67	5	2	5	1	3	1	5
P68	5	2	3	5	1	5	2

Table A1. *Cont.*

	Cutting(1)	Sanding(2)	Grinding(3)	Tempering(4)	Laminated(5)	Double Glazing(6)	Shipment(7)
P69	5	2	3	1	5	1	5
P70	5	2	2	5	1	3	5
P71	1	5	3	2	3	3	1
P72	3	3	5	5	1	3	5
P73	5	1	2	2	5	5	3
P74	5	3	5	2	1	1	5
P75	5	3	3	1	1	5	5
P76	5	2	2	5	5	1	1
P77	5	2	5	5	5	3	1
P78	3	2	3	1	5	5	5
P79	5	2	5	2	5	5	1
P80	5	1	2	5	5	1	2

Appendix B

Table A2. Monthly chart of factory personnel.

	Monday	Tuesday	Wednesday	Thursday	Friday	Saturday	Sunday	Day	Night
P1	1(1)	X	1(1)	5(2)	1(2)	1(2)	7(2)	14	11
	X	1(1)	1(1)	1(1)	1(2)	2(2)	3(2)		
	X	7(1)	3(1)	7(1)	1(1)	1(2)	1(2)		
	X	1(1)	1(1)	1(1)	1(1)	1(2)	1(2)		
	X	1(1)							
P2	X	4(1)	4(1)	2(1)	4(1)	4(1)	4(2)	12	13
	X	4(2)	4(2)	4(2)	7(2)	1(2)	4(2)		
	X	7(2)	4(2)	4(2)	2(2)	4(2)	4(2)		
	X	4(1)	4(1)	1(1)	4(1)	4(1)	X		
	2(1)	4(1)							
P3	X	5(2)	1(2)	6(2)	6(2)	7(2)	6(2)	14	10
	X	3(1)	2(1)	2(1)	1(1)	1(1)	3(1)		
	X	3(1)	6(1)	2(1)	6(2)	2(2)	X		
	3(1)	6(1)	6(1)	2(1)	3(2)	6(2)	X		
	X	3(1)							
P4	X	2(2)	1(2)	5(2)	2(2)	1(2)	5(2)	11	14
	X	2(1)	2(1)	6(1)	2(1)	2(2)	2(2)		
	X	2(1)	2(1)	2(1)	2(1)	2(1)	2(1)		
	X	2(2)	7(2)	2(2)	2(2)	2(2)	2(2)		
	X	2(1)							
P5	4(1)	4(1)	4(1)	4(1)	X	4(1)	4(2)	13	13
	4(2)	7(2)	5(2)	1(2)	X	6(1)	4(1)		
	6(1)	4(1)	4(2)	4(2)	X	4(2)	4(2)		
	4(2)	4(2)	1(2)	4(2)	X	7(1)	4(1)		
	4(1)	4(1)							
P6	5(1)	5(2)	5(2)	5(2)	5(2)	5(2)	X	13	12
	5(1)	4(1)	5(1)	5(1)	5(2)	5(2)	X		
	5(1)	5(1)	5(1)	1(2)	1(2)	5(2)	X		
	5(1)	5(1)	5(1)	5(1)	5(2)	X	5(1)		
	5(2)	X							
P7	X	2(2)	3(2)	1(2)	2(2)	2(2)	3(2)	13	12
	X	2(1)	2(1)	3(1)	3(1)	7(1)	2(1)		
	X	5(2)	3(2)	2(2)	4(2)	1(2)	2(2)		
	X	3(1)	3(1)	3(1)	2(1)	3(1)	3(1)		
	X	3(1)							

Table A2. *Cont.*

	Monday	Tuesday	Wednesday	Thursday	Friday	Saturday	Sunday	Day	Night
P8	X	5(1)	6(1)	6(1)	6(1)	5(2)	6(2)	13	12
	X	6(1)	5(1)	6(1)	6(1)	6(2)	6(2)		
	X	6(2)	6(2)	6(2)	6(2)	6(2)	6(2)		
	X	3(1)	6(1)	6(1)	6(2)	6(2)	X		
	6(1)	6(1)							
P9	2(2)	5(2)	X	5(2)	5(2)	5(2)	5(2)	12	14
	5(2)	2(2)	X	2(1)	5(1)	5(1)	2(1)		
	3(1)	2(2)	X	5(1)	5(1)	5(1)	3(1)		
	3(1)	4(2)	X	5(1)	5(1)	5(2)	5(2)		
	5(2)	5(2)							
P10	7(1)	2(1)	3(2)	5(2)	X	5(1)	5(1)	13	13
	5(1)	5(1)	5(2)	X	5(1)	5(2)	5(2)		
	5(2)	4(2)	5(2)	X	5(1)	5(1)	5(1)		
	5(2)	5(2)	1(2)	X	5(1)	5(1)	7(1)		
	5(2)	7(2)							
P11	2(1)	2(1)	2(1)	6(1)	2(1)	X	2(1)	14	12
	2(2)	2(2)	2(2)	2(2)	2(2)	X	2(2)		
	2(2)	2(2)	2(2)	2(2)	X	3(1)	1(1)		
	2(1)	2(1)	2(1)	2(2)	X	5(1)	2(1)		
	2(1)	5(2)							
P12	3(2)	X	2(1)	2(1)	2(1)	7(1)	2(2)	13	12
	2(2)	X	3(1)	3(1)	2(2)	3(2)	2(2)		
	2(2)	X	3(1)	2(1)	2(1)	3(1)	2(1)		
	4(2)	X	7(1)	2(1)	3(1)	2(2)	3(2)		
	2(2)	X							
P13	X	5(1)	3(1)	1(1)	1(1)	1(1)	1(2)	12	13
	X	1(2)	6(2)	2(2)	4(2)	6(2)	1(2)		
	X	1(1)	1(1)	1(1)	1(1)	2(2)	1(2)		
	X	3(1)	2(1)	1(2)	1(1)	1(2)	1(2)		
	X	1(1)							
P14	4(2)	3(2)	X	2(1)	3(1)	4(1)	3(1)	12	14
	4(2)	2(2)	X	2(1)	4(1)	4(1)	4(2)		
	4(2)	6(2)	X	4(1)	2(1)	4(1)	4(1)		
	2(2)	3(2)	X	2(1)	4(2)	1(2)	2(2)		
	4(2)	4(2)							
P15	5(2)	5(2)	5(2)	3(2)	3(2)	3(2)	X	12	14
	6(1)	5(1)	5(2)	5(2)	5(2)	5(2)	X		
	5(1)	1(1)	5(1)	5(1)	5(1)	X	5(1)		
	5(1)	5(2)	5(2)	5(2)	5(2)	X	5(1)		
	5(1)	5(1)							
P16	7(2)	3(2)	7(2)	X	2(1)	3(1)	7(1)	14	12
	7(1)	7(1)	7(2)	X	2(1)	4(1)	1(1)		
	7(1)	7(1)	7(1)	X	7(1)	7(1)	7(2)		
	7(2)	1(2)	X	7(1)	7(2)	7(2)	7(2)		
	7(2)	2(2)							
P17	5(1)	3(1)	5(1)	5(1)	X	5(1)	5(1)	12	14
	4(1)	5(2)	5(2)	5(2)	X	5(1)	5(1)		
	5(1)	5(1)	5(2)	5(2)	X	1(1)	5(2)		
	2(2)	5(2)	3(2)	5(2)	X	5(2)	5(2)		
	1(2)	5(2)							
P18	6(2)	6(2)	6(2)	X	6(1)	6(2)	3(2)	14	11
	6(2)	6(2)	X	6(1)	6(1)	6(1)	6(1)		
	6(1)	X	6(1)	6(1)	6(1)	4(1)	6(2)		
	6(2)	X	6(1)	6(1)	6(1)	6(1)	7(2)		
	7(2)	X							

Table A2. *Cont.*

	Monday	Tuesday	Wednesday	Thursday	Friday	Saturday	Sunday	Day	Night
P19	3(1)	7(1)	7(1)	7(1)	7(2)	7(2)	X	14	12
	7(1)	7(1)	7(1)	7(1)	7(1)	7(2)	X		
	7(2)	3(2)	7(2)	3(2)	3(2)	7(2)	X		
	7(1)	7(1)	7(1)	7(1)	7(1)	7(2)	X		
	7(2)	7(2)							
P20	6(1)	7(1)	7(1)	7(1)	7(2)	6(2)	X	14	12
	3(2)	6(2)	7(2)	6(2)	6(2)	X	7(1)		
	6(2)	6(2)	6(2)	X	6(1)	7(1)	7(1)		
	2(1)	3(1)	3(1)	X	6(1)	7(1)	7(1)		
	7(2)	7(2)							
P21	3(1)	4(1)	3(1)	4(2)	X	3(1)	2(1)	14	12
	3(1)	3(2)	3(2)	3(2)	X	3(1)	3(1)		
	3(1)	3(1)	3(2)	3(2)	X	3(2)	3(2)		
	3(2)	3(2)	3(2)	3(2)	X	3(1)	3(1)		
	3(1)	3(1)							
P22	X	5(1)	4(1)	1(1)	1(1)	7(2)	1(2)	14	11
	X	1(2)	3(2)	5(2)	1(2)	3(2)	1(2)		
	X	6(1)	2(1)	1(1)	2(2)	4(2)	3(2)		
	X	1(1)	7(1)	7(1)	3(1)	1(1)	X		
	3(1)	3(1)							
P23	4(2)	4(2)	X	4(1)	4(1)	2(1)	4(1)	12	13
	2(1)	4(2)	X	2(2)	4(2)	2(2)	4(2)		
	X	4(1)	3(1)	3(1)	4(1)	2(1)	2(2)		
	X	2(1)	4(1)	4(2)	3(2)	4(2)	3(2)		
	X	2(2)							
P24	6(1)	6(1)	X	2(1)	6(1)	6(1)	6(1)	14	11
	6(2)	X	6(1)	6(1)	6(1)	6(2)	6(2)		
	6(2)	X	4(1)	6(1)	6(1)	6(1)	6(2)		
	6(2)	X	6(1)	6(2)	6(2)	6(2)	6(2)		
	6(2)	X							
P25	7(2)	7(2)	7(2)	7(2)	X	7(1)	7(1)	13	13
	7(1)	7(1)	4(1)	7(1)	X	7(2)	7(2)		
	7(2)	7(2)	7(2)	7(2)	X	3(1)	7(1)		
	3(2)	7(2)	3(2)	X	7(1)	7(1)	2(1)		
	7(1)	7(1)							
P26	2(2)	X	2(1)	2(2)	2(2)	2(2)	2(2)	12	13
	2(2)	X	4(1)	7(1)	2(1)	2(1)	5(1)		
	4(1)	X	1(1)	5(1)	1(1)	3(2)	7(2)		
	7(2)	X	5(1)	2(1)	2(2)	2(2)	2(2)		
	2(2)	X							
P27	3(2)	X	X	3(1)	3(1)	6(1)	3(1)	11	14
	3(2)	3(2)	X	4(1)	6(2)	3(2)	3(2)		
	3(2)	3(2)	X	1(1)	3(1)	2(1)	1(1)		
	3(1)	5(1)	X	3(2)	3(2)	3(2)	3(2)		
	3(2)	3(2)							
P28	5(1)	5(1)	5(1)	5(1)	5(1)	X	4(2)	14	12
	4(2)	5(2)	5(2)	7(2)	5(2)	X	4(1)		
	4(1)	7(1)	5(2)	5(2)	5(2)	X	5(1)		
	1(1)	5(1)	5(2)	7(2)	5(2)	X	5(1)		
	1(1)	7(1)							
P29	6(1)	2(1)	6(1)	6(1)	3(1)	5(1)	X	13	13
	6(2)	6(2)	6(2)	6(2)	4(2)	6(2)	X		
	6(2)	6(2)	6(2)	6(2)	6(2)	6(2)	X		
	6(1)	6(1)	6(1)	6(1)	6(2)	X	4(1)		
	4(1)	6(1)							

Table A2. *Cont.*

	Monday	Tuesday	Wednesday	Thursday	Friday	Saturday	Sunday	Day	Night
P30	3(1)	3(1)	X	4(1)	7(1)	3(2)	3(2)	11	14
	7(2)	3(2)	X	1(1)	4(1)	3(1)	3(2)		
	3(2)	3(2)	X	7(1)	3(1)	7(2)	7(2)		
	3(2)	4(2)	X	3(1)	3(1)	6(2)	3(2)		
	3(2)	X							
P31	5(2)	5(2)	5(2)	5(2)	5(2)	X	5(1)	12	14
	3(1)	5(1)	4(2)	5(2)	5(2)	X	5(1)		
	5(1)	5(1)	5(1)	5(1)	5(1)	X	5(1)		
	5(1)	5(1)	6(2)	5(2)	X	3(2)	5(2)		
	3(2)	1(2)							
P32	1(2)	1(2)	1(2)	7(2)	1(2)	X	1(1)	12	14
	1(1)	1(1)	1(1)	X	4(1)	5(1)	7(1)		
	1(1)	7(1)	1(1)	X	7(1)	1(1)	6(2)		
	1(2)	1(2)	7(2)	X	7(2)	7(2)	7(2)		
	4(2)	7(2)							
P33	5(2)	6(2)	6(2)	X	6(2)	6(2)	2(2)	14	12
	6(2)	6(2)	6(2)	X	6(1)	6(1)	6(1)		
	6(1)	3(2)	6(2)	X	6(1)	6(1)	6(1)		
	6(1)	6(1)	6(2)	X	6(1)	4(1)	6(1)		
	6(1)	6(1)							
P34	6(2)	7(2)	X	3(1)	3(2)	3(2)	6(2)	12	13
	1(2)	3(2)	X	6(1)	5(1)	6(1)	3(1)		
	3(1)	2(1)	X	6(1)	6(1)	6(1)	3(1)		
	3(1)	X	3(2)	6(2)	3(2)	3(2)	4(2)		
	4(2)	X							
P35	7(2)	5(2)	7(2)	2(2)	7(2)	7(2)	X	14	12
	7(1)	7(1)	7(1)	7(1)	7(1)	7(2)	X		
	7(1)	5(1)	7(1)	5(1)	7(2)	7(2)	X		
	7(1)	7(1)	7(1)	7(2)	7(2)	5(2)	X		
	7(1)	7(1)							
P36	6(1)	6(1)	6(2)	6(2)	6(2)	X	6(1)	13	13
	6(1)	6(1)	6(1)	6(1)	7(1)	X	6(1)		
	6(1)	6(1)	2(2)	6(2)	1(2)	X	6(1)		
	2(2)	6(2)	1(2)	6(2)	1(2)	X	6(1)		
	6(2)	6(2)							
P37	X	7(1)	5(1)	4(2)	4(2)	4(2)	2(2)	11	14
	X	4(1)	7(1)	5(2)	7(2)	4(2)	X		
	2(1)	4(2)	7(2)	2(2)	2(2)	7(2)	X		
	7(1)	2(1)	2(1)	4(1)	4(1)	6(1)	X		
	2(2)	4(2)							
P38	5(1)	1(1)	1(1)	1(2)	X	1(1)	6(1)	13	13
	1(1)	1(1)	1(2)	1(2)	X	1(1)	1(1)		
	1(1)	1(2)	4(2)	2(2)	X	4(1)	1(1)		
	1(1)	1(2)	7(2)	1(2)	X	4(2)	1(2)		
	6(2)	1(2)							
P39	6(2)	7(2)	X	5(1)	6(1)	3(1)	2(1)	12	13
	2(1)	X	2(1)	2(2)	2(2)	5(2)	5(2)		
	2(2)	X	2(1)	2(1)	5(1)	5(1)	3(1)		
	X	2(2)	2(2)	2(2)	5(2)	5(2)	2(2)		
	X	5(1)							
P40	7(1)	2(2)	2(2)	7(2)	7(2)	X	4(1)	13	13
	5(1)	5(1)	7(1)	7(2)	7(2)	X	7(1)		
	7(1)	7(2)	3(2)	7(2)	7(2)	X	7(1)		
	4(1)	7(2)	7(2)	7(2)	X	3(1)	7(1)		
	7(1)	7(1)							

Table A2. Cont.

	Monday	Tuesday	Wednesday	Thursday	Friday	Saturday	Sunday	Day	Night
P41	6(1)	6(1)	6(1)	6(1)	6(1)	X	6(1)	14	12
	6(1)	6(1)	6(1)	6(2)	3(2)	X	6(1)		
	6(1)	6(2)	6(2)	6(2)	6(2)	X	6(2)		
	6(2)	6(2)	2(2)	6(2)	6(2)	X	6(1)		
	6(1)	6(1)							
P42	4(2)	3(2)	3(2)	X	4(1)	2(1)	4(2)	12	13
	4(2)	3(2)	X	5(1)	3(1)	4(1)	4(1)		
	4(2)	4(2)	X	3(2)	4(2)	5(2)	3(2)		
	4(2)	X	5(1)	4(1)	2(1)	4(1)	4(1)		
	3(1)	X							
P43	2(2)	6(2)	2(2)	X	7(1)	2(1)	7(1)	12	14
	2(1)	2(1)	6(2)	X	6(1)	2(1)	7(2)		
	2(2)	2(2)	2(2)	X	7(1)	7(1)	5(2)		
	7(2)	2(2)	2(2)	X	6(1)	2(1)	6(1)		
	6(2)	2(2)							
P44	6(1)	2(1)	3(1)	6(2)	X	6(1)	3(1)	14	12
	3(1)	3(1)	3(1)	5(2)	X	2(1)	2(2)		
	3(2)	2(2)	6(2)	3(2)	X	6(1)	6(1)		
	6(1)	2(1)	3(1)	2(2)	X	2(2)	6(2)		
	3(2)	3(2)							
P45	6(2)	6(2)	X	3(1)	6(1)	6(1)	6(1)	13	13
	6(1)	6(2)	X	6(1)	3(2)	6(2)	6(2)		
	6(2)	6(2)	X	6(1)	2(1)	6(1)	6(1)		
	6(2)	6(2)	X	6(1)	6(1)	6(1)	6(2)		
	6(2)	6(2)							
P46	3(1)	3(1)	3(1)	3(1)	3(2)	X	6(1)	13	13
	3(2)	5(2)	3(2)	3(2)	3(2)	X	3(1)		
	2(1)	1(1)	3(1)	3(1)	3(2)	X	3(1)		
	5(2)	3(2)	5(2)	3(2)	2(2)	X	3(1)		
	5(1)	3(2)							
P47	6(2)	6(2)	6(2)	6(2)	X	6(1)	6(1)	14	12
	6(1)	3(1)	6(1)	2(1)	X	6(1)	6(2)		
	6(2)	6(2)	6(2)	6(2)	X	6(1)	5(1)		
	6(1)	6(1)	6(1)	6(1)	X	6(1)	6(2)		
	6(2)	6(2)							
P48	5(2)	6(2)	5(2)	3(2)	5(2)	X	5(1)	12	14
	5(2)	5(2)	4(2)	4(2)	X	5(1)	5(1)		
	5(1)	5(1)	5(1)	5(2)	X	5(1)	4(1)		
	5(1)	5(1)	5(2)	5(2)	X	2(1)	5(1)		
	5(2)	5(2)							
P49	4(2)	4(2)	4(2)	2(2)	4(2)	4(2)	X	13	13
	6(1)	4(1)	4(1)	4(1)	5(1)	2(1)	X		
	4(1)	4(1)	4(2)	4(2)	5(2)	2(2)	X		
	4(1)	4(1)	4(2)	5(2)	4(2)	X	2(1)		
	3(1)	5(1)							
P50	X	6(1)	6(2)	6(2)	6(2)	6(2)	6(2)	11	14
	X	6(1)	2(2)	6(2)	6(2)	6(2)	6(2)		
	X	3(1)	6(1)	6(1)	6(2)	6(2)	X		
	6(1)	6(1)	6(1)	6(1)	6(2)	4(2)	X		
	1(1)	6(1)							
P51	X	4(2)	5(2)	6(2)	4(2)	4(2)	1(2)	14	10
	X	6(1)	4(1)	4(1)	4(1)	4(2)	4(2)		
	X	2(1)	4(1)	4(2)	4(2)	X	2(1)		
	2(1)	4(1)	4(1)	4(1)	4(1)	X	X		
	4(1)	4(1)							

Table A2. Cont.

	Monday	Tuesday	Wednesday	Thursday	Friday	Saturday	Sunday	Day	Night
P52	2(1)	4(1)	6(1)	X	6(2)	2(2)	6(2)	14	12
	2(2)	2(2)	6(2)	X	6(1)	6(1)	2(1)		
	2(1)	2(1)	6(1)	X	6(1)	6(2)	2(2)		
	6(2)	2(2)	2(2)	X	2(1)	2(1)	2(1)		
	6(1)	2(2)							
P53	6(2)	X	6(1)	6(1)	1(1)	6(1)	3(1)	13	12
	6(2)	X	6(2)	6(2)	6(2)	6(2)	6(2)		
	X	5(1)	6(1)	6(1)	4(2)	6(2)	6(2)		
	X	6(1)	5(1)	6(1)	6(1)	6(1)	6(2)		
	X	6(2)							
P54	2(1)	5(1)	5(1)	5(1)	5(2)	5(2)	X	14	12
	5(1)	5(1)	5(1)	5(1)	5(2)	5(2)	X		
	1(1)	5(2)	5(2)	5(2)	5(2)	5(2)	X		
	5(1)	5(1)	X	5(1)	5(1)	5(1)	5(2)		
	5(2)	5(2)							
P55	5(1)	7(1)	7(1)	7(2)	X	4(1)	7(1)	13	13
	7(1)	7(1)	5(1)	7(2)	X	7(1)	7(1)		
	7(2)	5(2)	7(2)	7(2)	X	7(1)	1(2)		
	1(2)	3(2)	4(2)	X	6(1)	5(1)	7(2)		
	7(2)	7(2)							
P56	4(1)	6(1)	4(2)	4(2)	X	7(1)	4(1)	12	14
	4(1)	4(2)	4(2)	4(2)	X	6(1)	4(1)		
	4(2)	4(2)	2(2)	X	4(1)	2(1)	4(1)		
	4(1)	4(1)	X	4(2)	4(2)	7(2)	4(2)		
	4(2)	5(2)							
P57	3(2)	3(2)	4(2)	2(2)	X	3(2)	3(2)	11	14
	5(2)	X	3(1)	3(1)	2(1)	3(1)	5(1)		
	6(2)	X	5(1)	3(1)	3(2)	3(2)	3(2)		
	X	6(1)	3(1)	3(1)	2(2)	3(2)	5(2)		
	X	2(1)							
P58	3(2)	1(2)	2(2)	X	3(2)	2(2)	5(2)	14	12
	3(2)	5(2)	2(2)	X	3(1)	7(1)	2(1)		
	2(1)	3(1)	3(2)	X	3(1)	3(1)	7(1)		
	2(1)	7(1)	5(2)	X	3(1)	3(1)	1(1)		
	1(1)	3(2)							
P59	1(1)	1(1)	2(1)	1(1)	5(1)	X	5(2)	11	13
	1(2)	X	1(2)	1(2)	1(2)	1(2)	1(2)		
	1(2)	X	1(2)	1(2)	1(2)	1(2)	5(2)		
	1(2)	X	1(1)	1(1)	1(1)	1(1)	1(1)		
	X	2(1)							
P60	7(1)	7(1)	7(1)	7(1)	7(1)	7(1)	X	12	13
	7(2)	7(2)	7(2)	7(2)	X	1(1)	X		
	7(1)	7(1)	7(2)	7(2)	7(2)	X	7(2)		
	7(2)	7(2)	7(2)	7(2)	7(2)	X	7(1)		
	7(1)	1(1)							
P61	1(2)	X	1(1)	1(2)	1(2)	1(2)	6(2)	12	13
	1(2)	X	1(1)	1(1)	1(1)	X	1(1)		
	1(2)	1(2)	1(2)	1(2)	X	1(1)	5(1)		
	5(1)	1(1)	1(1)	X	1(1)	1(1)	6(2)		
	1(2)	1(2)							
P62	4(1)	5(1)	5(1)	X	5(1)	5(1)	5(1)	12	14
	5(1)	5(1)	5(1)	X	5(1)	6(1)	5(1)		
	5(2)	5(2)	5(2)	X	5(2)	5(2)	5(2)		
	5(2)	5(2)	5(2)	X	5(2)	5(2)	5(2)		
	5(2)	6(2)							

Table A2. *Cont.*

	Monday	Tuesday	Wednesday	Thursday	Friday	Saturday	Sunday	Day	Night
P63	4(1)	X	4(1)	4(1)	4(1)	1(1)	4(1)	13	12
	4(1)	X	5(2)	4(2)	4(2)	4(2)	5(2)		
	4(2)	X	2(1)	6(1)	4(1)	4(2)	4(2)		
	4(2)	X	4(1)	4(1)	4(1)	4(2)	4(2)		
	X	4(2)							
P64	6(2)	X	6(1)	7(1)	6(1)	6(2)	7(2)	13	12
	6(2)	X	6(1)	6(1)	1(1)	3(1)	6(2)		
	6(2)	X	6(2)	6(2)	6(2)	6(2)	6(2)		
	6(2)	X	6(1)	6(1)	5(1)	6(1)	1(1)		
	6(1)	X							
P65	7(1)	X	7(1)	3(2)	7(2)	7(2)	7(2)	11	14
	7(2)	X	7(1)	3(2)	7(2)	7(2)	6(2)		
	5(2)	X	7(1)	7(1)	7(2)	6(2)	7(2)		
	X	7(1)	7(1)	7(1)	7(1)	7(1)	7(2)		
	X	7(1)							
P66	1(1)	3(1)	3(2)	X	3(1)	3(1)	2(1)	14	11
	3(1)	3(1)	3(2)	X	3(2)	2(2)	3(2)		
	3(2)	6(2)	X	3(1)	3(2)	3(2)	4(2)		
	7(2)	X	2(1)	3(1)	6(1)	6(1)	3(1)		
	2(1)	X							
P67	1(2)	X	6(1)	6(1)	6(1)	6(1)	1(1)	14	11
	X	6(1)	6(1)	3(1)	6(2)	6(2)	X		
	6(1)	6(1)	6(1)	6(2)	6(2)	6(2)	X		
	6(1)	6(2)	6(2)	6(2)	6(2)	6(2)	X		
	6(1)	6(1)							
P68	2(2)	7(2)	X	5(1)	5(1)	5(1)	7(2)	14	12
	5(2)	5(2)	X	5(1)	5(2)	3(2)	5(2)		
	5(2)	5(2)	X	5(1)	7(1)	5(1)	2(1)		
	5(2)	5(2)	X	5(1)	5(1)	5(1)	5(1)		
	5(1)	5(1)							
P69	6(1)	6(1)	X	6(1)	6(2)	6(2)	6(2)	11	14
	6(2)	1(2)	X	3(1)	6(1)	4(1)	6(1)		
	6(2)	X	4(1)	4(1)	6(1)	6(1)	6(2)		
	6(2)	X	6(2)	6(2)	6(2)	6(2)	4(2)		
	6(2)	X							
P70	5(1)	6(2)	5(2)	X	5(1)	5(1)	5(2)	13	13
	5(2)	6(2)	1(2)	X	6(1)	5(1)	5(2)		
	5(2)	5(2)	5(2)	X	5(1)	5(1)	5(2)		
	5(2)	5(2)	X	5(1)	5(1)	4(1)	6(1)		
	5(1)	5(1)							
P71	7(1)	1(1)	7(2)	X	7(1)	6(1)	5(1)	12	14
	1(1)	7(2)	7(2)	X	7(1)	1(2)	7(2)		
	1(2)	1(2)	X	7(1)	7(1)	4(1)	6(1)		
	6(1)	7(2)	X	1(2)	1(2)	5(2)	6(2)		
	1(2)	6(2)							
P72	5(2)	2(2)	6(2)	3(2)	4(2)	5(2)	X	13	13
	5(1)	2(1)	6(1)	5(1)	5(1)	5(2)	X		
	5(1)	6(1)	5(1)	5(2)	5(2)	5(2)	X		
	1(1)	6(1)	5(1)	5(2)	5(2)	6(2)	X		
	5(1)	5(1)							
P73	2(1)	1(2)	2(2)	X	2(1)	2(1)	7(1)	12	14
	2(1)	6(2)	2(2)	X	2(2)	7(2)	7(2)		
	7(2)	7(2)	X	4(1)	2(2)	2(2)	2(2)		
	2(2)	6(2)	X	7(1)	2(1)	2(1)	6(1)		
	2(1)	6(1)							

Table A2. *Cont.*

	Monday	Tuesday	Wednesday	Thursday	Friday	Saturday	Sunday	Day	Night
P74	6(1)	6(1)	6(1)	6(1)	5(1)	X	6(2)	12	14
	6(2)	6(2)	6(2)	6(2)	6(2)	X	6(1)		
	6(1)	6(1)	6(1)	6(1)	X	6(1)	6(1)		
	3(2)	6(2)	6(2)	6(2)	X	6(2)	6(2)		
	6(2)	6(2)							
P75	5(2)	X	5(1)	5(1)	5(2)	5(2)	5(2)	11	13
	5(2)	X	5(1)	5(1)	3(1)	5(1)	5(2)		
	5(2)	X	1(2)	5(2)	5(2)	5(2)	5(2)		
	5(2)	X	5(1)	5(1)	7(1)	5(1)	X		
	5(1)	X							
P76	X	6(1)	6(2)	6(2)	6(2)	6(2)	X	14	11
	6(1)	6(1)	6(1)	6(2)	7(2)	X	6(1)		
	6(1)	6(1)	7(1)	7(1)	6(1)	X	6(1)		
	6(2)	6(2)	6(2)	6(2)	6(2)	X	6(1)		
	6(1)	6(1)							
P77	7(2)	X	7(2)	7(2)	X	6(2)	7(2)	12	13
	7(2)	7(2)	7(2)	7(2)	X	7(1)	7(1)		
	7(1)	6(1)	7(1)	X	3(1)	7(1)	7(1)		
	7(1)	7(2)	6(2)	X	7(1)	7(1)	7(1)		
	2(2)	4(2)							
P78	X	4(2)	4(2)	4(2)	2(2)	4(2)	X	14	11
	4(1)	4(1)	3(1)	4(1)	6(2)	4(2)	X		
	4(1)	4(1)	4(1)	4(1)	4(1)	X	4(1)		
	4(1)	4(2)	4(2)	4(2)	4(2)	X	4(1)		
	4(1)	4(1)							
P79	7(2)	7(2)	X	7(1)	7(1)	7(1)	1(1)	11	14
	7(2)	7(2)	X	7(1)	7(1)	7(1)	7(2)		
	7(2)	7(2)	X	7(2)	7(2)	7(2)	X		
	7(1)	7(1)	4(2)	7(2)	7(2)	7(2)	X		
	7(1)	2(1)							
P80	6(2)	6(2)	6(2)	6(2)	6(2)	X	6(1)	12	14
	6(1)	6(1)	6(2)	6(2)	6(2)	X	6(1)		
	3(1)	6(1)	6(1)	6(2)	6(2)	X	6(1)		
	6(1)	6(2)	6(2)	3(2)	X	6(1)	6(1)		
	6(1)	6(2)							

Notes: $l(k)$: l: shift; k: section assigned personnel.

References

1. Eren, T.; Varli, E. Shift Scheduling Problems and A Case Study. *Int. J. Inform. Technol.* **2017**, *10*, 185–197.
2. Sungur, B. Integer Programming Model for Fuzzy Shift Scheduling Problems. *Erciyes Univ. J. Fac. Econ. Adm. Sci.* **2008**, *30*, 211–227.
3. Varli, E. Solution of the Shift Scheduling Problem for the Foremans in the Manufacturing Sector with Ahp-Goal Programming. Master's Thesis, Kirikkale University, Kirikkale, Turkey, 2017.
4. Özcan, E.C.; Varli, E.; Eren, T. Goal Programming Approach for Shift Scheduling Problems in Hydroelectric Power Plants. *J. Inf. Technol.* **2017**, *10*, 363–370.
5. Güngör, İ. A Model Proposal for Nursing Assignment and Scheduling. *Süleyman Demirel Univ. J. Fac. Econ. Adm. Sci.* **2002**, *7*, 77–94.
6. Bard, J.F.; Binici, C. Staff Scheduling at the United States Postal Service. *Comput. Oper. Res.* **2003**, *30*, 745–771. [CrossRef]
7. Wong, G.Y.C.; Chun, A.H.W. Constraint-Based Rostering Using Meta-Level Reasoning and Probability-Based Ordering. *Eng. Appl. Artif. Intell.* **2004**, *17*, 599–610. [CrossRef]

8. Azaiez, M.N.; Al Sharif, S.S. A 0-1 Goal Programming Model for Nurse Scheduling. *Comput. Oper. Res.* **2005**, *32*, 491–507. [CrossRef]
9. Seçkiner, S.U.; Gökçen, H.; Kurt, M. An Integer Programming Model for Hierarchical Workforce Scheduling Problem. *Eur. J. Oper. Res.* **2007**, *183*, 694–699. [CrossRef]
10. Castillo, I.; Joro, T.; Li, Y.Y. Workforce Scheduling with Multiple Objectives. *Eur. J. Oper. Res.* **2009**, *196*, 162–170. [CrossRef]
11. Olive, J.T. A Proposed Mathematical Model for the Personnel Scheduling Problem in a Manufacturing Company. Master's Thesis, Istanbul Technical University, İstanbul, Turkey, 2009.
12. Topaloğlu, S. A Shift Scheduling Model for Employees with Different Seniority Levels and an Application in Healthcare. *Eur. J. Oper. Res.* **2009**, *198*, 943–957. [CrossRef]
13. Heimerl, C.; Kolisch, R. Scheduling and Staffing Multiple Projects with a Multi-Skilled Workforce. *OR Spectr.* **2010**, *32*, 343–368. [CrossRef]
14. Karaatli, M. Multi Purpose Manpower Scheduling in Fuzzy Environment: An Application for Nurses. Ph.D. Thesis, Süleyman Demirel University, Isparta, Turkey, 2010.
15. Koruca, H.İ. Development of A Simulation-Based Shift Planning Module. *J. Fac. Eng. Archit. Gazi Univ.* **2010**, *25*, 469–482.
16. Atmaca, E.; Pehlivan, C.; Aydoğdu, B.; Yakici, M. Nurse Scheduling Problem and Application. *Erciyes Univ. J. Inst. Sci. Technol.* **2012**, *28*, 351–358.
17. Bağ, N.; Özdemir, N.M.; Eren, T. 0-1 Goal Programming and ANP Method and Nurse Scheduling Problem Solution. *Int. J. Eng. Res. Dev.* **2012**, *4*, 2–6.
18. Bektur, G.; Hasgül, S. Workforce Scheduling Problem According to Seniority Levels: An Application For A Service System. *Afyon Kocatepe Univ. J. Fac. Econ. Adm. Sci.* **2013**, *15*, 385–402.
19. Çöl, G. Goal Programming Approach to Workforce Scheduling Problem with Different Seniority Levels. Master's Thesis, Eskişehir Osmangazi University, Eskişehir, Turkey, 2013.
20. Ünal, F.M.; Eren, T. The Solution of Shift Scheduling Problem by Using Goal Programming. *Acad. Platf. J. Eng. Sci.* **2016**, *4*, 28–37.
21. Özder, E.H.; Varli, E.; Eren, T. A Model Suggestion for Cleaning Staff Scheduling Problem with Goal Programming Approach. *Karadeniz J. Sci.* **2017**, *7*, 114–127.
22. Varli, E.; Alağaş, H.M.; Eren, T.; Özder, E.H. Goal Programming Solution of the Examiner Assignment Problem. *Bilge Int. J. Sci. Technol. Res.* **2017**, *1*, 105–118.
23. Varli, E.; Ergişi, B.; Eren, T. Nurse Scheduling Problem with Special Constraints: Goal Programming Approach. *Erciyes Univ. J. Fac. Econ. Adm. Sci.* **2017**, *49*, 189–206.
24. Bedir, N. Shift Scheduling Problems Solution with Combined Ahppromethee and Goal Programming Methods: An Aplication in Hydroelectric Power Plant. Master's Thesis, Kirikkale University, Kirikkale, Turkey, 2018.
25. Gür, Ş.; Eren, T. Scheduling and Planning in Service Systems with Goal Programming: Literature Review. *Mathematics* **2018**, *6*, 265. [CrossRef]
26. Koç, E. An Exact Approach for a Dynamic Workforce Scheduling Problem. Master's Thesis, Özyeğin University, İstanbul, Turkey, 2018.
27. Koçtepe, S.; Bedir, N.; Eren, T.; Gür, Ş. Solution of Personnel Scheduling Problem for Organization Officers With 0-1 Integer Programming. *J. Econ. Bus. Manag.* **2018**, *2*, 25–46.
28. Tapkan, P.Z.; Özbakir, L.; Kulluk, S.; Telcioğlu, B. Modelling and solving railway crew rostering problem. *J. Fac. Eng. Archit. Gazi Univ.* **2018**, *33*, 953–965.
29. Özder, E.H.; Özcan, E.; Eren, T. Staff Task-Based Shift Scheduling Solution with an ANP and Goal Programming Method in a Natural Gas Combined Cycle Power Plant. *Mathematics* **2019**, *7*, 192. [CrossRef]

 © 2019 by the authors. Licensee MDPI, Basel, Switzerland. This article is an open access article distributed under the terms and conditions of the Creative Commons Attribution (CC BY) license (http://creativecommons.org/licenses/by/4.0/).

Article

Evaluation of Elite Athletes Training Management Efficiency Based on Multiple Criteria Measure of Conditioning Using Fewer Data

Aleksandras Krylovas [1], Natalja Kosareva [1], Rūta Dadelienė [2] and Stanislav Dadelo [3,*]

[1] Department of Mathematical Modelling, Vilnius Gediminas Technical University, Saulėtekio al. 11, 10221 Vilnius, Lithuania; aleksandras.krylovas@vgtu.lt (A.K.); natalja.kosareva@vgtu.lt (N.K.)
[2] Department of Rehabilitation, Physical and Sports Medicine, Institute of Health Science, Vilnius University, Saulėtekio al. 11, 10221 Vilnius, Lithuania; ruta.dadeliene@mf.vu.lt
[3] Department of Entertainment Industries, Vilnius Gediminas Technical University, Saulėtekio al. 11, 10221 Vilnius, Lithuania
* Correspondence: stanislav.dadelo@vgtu.lt

Received: 10 December 2019; Accepted: 31 December 2019; Published: 2 January 2020

Abstract: Innovative solutions and techniques in the sports industry are commonly used and tested in real conditions. Elite athletes have to achieve their peak performance before the main competition of the year, which is the World Championship, and every fourth year before the Olympic Games, when the main competition of athletes takes place. The present study aims to analyze and evaluate the ability of elite kayakers to achieve the best form at the right times, with the Olympic Games taking the greatest importance. Target values for multiple measures of conditioning are compared to target values set by experts. A weighted least squares metric with weights varied by time period is developed as a measure of fulfillment of the athletes' conditioning plans. The novelty of the paper is the idea of using linear combination of polynomials and trigonometric functions for approximating the target functions and application of the proposed methodology for the optimization and evaluation of athletic training.

Keywords: athletes' condition; approximation; parameter estimation; least squares method; visualization

1. Introduction

Sports are a phenomenon of global importance. The investment in professional athletes is a particularly important process. Sporting events promote the development of powerful worldwide industries. A key factor in sports is the ability to develop training schedules to optimize the athletes' training conditions. The optimization and evaluation of athletic training is the most important problem for sports scientists, coaches, and athletes. Elite athletes are willing to perform more voluminous and high-quality training routines, and this process requires effective management. Elite athletes attain a world-class status in endurance sports after four to seven years of specialized training [1]. Over this period, they usually have 3000–7000 h of effective training and cover distances of 3000–4000 km/yr [2]. Winning medals in international competitions requires not only outstanding abilities and the long-term training of athletes, it is also crucial that they achieve peak performance at the right time.

The Olympic Games is the main competition for the Olympic sports athletes. There are four years of the Olympic training cycle, with its organizational structure and training methodology directed at making the successful start of athletes in the Olympic Games [3]. The elite athletes' training must be carefully planned for several years in the future and based on individual indicators (training load specification and the athlete's body adaptation to a particular training load). The process of the elite athletes' training is complex and difficult to predict because of a number of unknown factors (i.e., accidents, acclimatization, illnesses, psychological changes, the pace of recovery, etc.) [4].

The Olympic training cycle consists of four one-year macro cycles, with the main competition, the World Championship, taking place in each of these cycles. The athletes make strenuous efforts to be in their best form before the World Championship. In the last year of the Olympic cycle, some of the athletes get a chance to participate in the Olympic Games during the World Championship. It means that athletes have to achieve their peak performance twice a year. In this case, the management of training is very complicated, because the athletes need to achieve their peak performance in a short time [5,6].

Monitoring the training load and athletes' physiological adaptation is essential for optimizing training and minimizing the risk of overtraining, injuries, illnesses, etc. Usually, sports managers and scientists are restricted in publishing the data on the athletes' condition before competitions. In elite sports, monitoring is extensive, but most of the data remain confidential [7].

Scientists are searching for novel approaches to visualize and analyze the data obtained in the training sessions. One of the most relevant research topics is visualization for motion analysis, which is important for training optimization, technical and tactical improvements, as well as the prevention of injuries [8]. However, not only motion optimization is required for achieving good results. The growing need causes the necessity for making a training process much more technologically advanced, and this is an important field of interest for athletes, coaches, and scientists. Massive amounts of data are generated and used in sports medicine, including preventative care and rehabilitation. Most major professional teams today make data-driven decisions and employ the analytical staff to help prepare training plans, predict athlete risks, and prescribe personalized recovery strategies. Most of the athletes use trackers to measure, accompany, and control the data (e.g., power, velocity, duration, altitude, and heart rate) obtained in training and later perform an individual analysis and online planning [9]. The present study focuses on the visualization of the training process based on the physiological condition of athletes.

The optimization of the athletic training process is an appropriate application of operational research (OR). Prediction management and decision making in athletic training should be based on objective data analysis and parameter estimation [10].

Reviewing the literature related to the application of mathematical models and methods used for solving similar problems has shown that the authors use various approaches. For example, Armstrong, Weidner, and Walker [11] have elaborated on the analysis of variance (ANOVA) and t-test of independent samples for clinical proficiency evaluation in athletic training. A qualitative analysis of the respondents' comments has also been performed.

Li, Zhu, Chen, and Xue [12] have elaborated on the balanced approach of data envelopment analysis (DEA) to cross-efficiency evaluation. An iterative algorithm for obtaining the final optimal and balanced cross-efficiency score has been developed by the authors.

Four imbalance indicators for measuring the difference between standard distribution and the ideally balanced distribution have been proposed by Karsu and Morton [13]. The interesting phenomenon is the total componentwise proportional deviation because it is an individual-oriented measure, which is the weighted sum of fractional misallocations for each party.

The Delphi method has been used by Reefke and Sundaram [14] to identify a set of key sustainable supply chain management (SSCM) problems and the associated research opportunities. The data components have been synthesized, and the parameters have been estimated according to their relative importance based on the experts' judgments. New insights into the potential dependencies between the factors and their influence on the success of SSCM have been provided in this work. The development of the schedule for the sporting activities of the Ecuadorian football federation under a set of constraints with the use of integer programming has been proposed by Recalde, Torres, and Vaca [15]. A heuristic three-phase approach has also been adjusted to solve the considered problem. The developed methodology provides more benefits than the empirical method.

Algorithms based on multiple criteria decision making are often used in solving various problems in sports management. Dadelo, Turskis, Zavadskas, and Dadelienė [16] have described a novel framework for practical assessment and the ranking of basketball players based on the adjusted well-known Technique for Order of Preference by Similarity to Ideal Solution (TOPSIS) method. The graphical visualization of similarities provides further insights [17,18]. Visualization solutions of processes are better understood, allowing one to intuitively select a more appropriate optimization method. The visualization of processes is particularly useful for the cases with a variety of large and small factors, which should be controlled [19]. This is particularly important for inexperienced (non-professional) users.

The present study aims to analyze the ability of elite kayakers to achieve peak performance at the right times, with the Olympic Games taking the greatest importance. Real data is collected from two Olympic kayakers, and regression analysis is used (a combination of polynomial and trigonometric functions) to model how well their conditioning fits the ideal. A weighted least squares metric with weights varied by time period is developed as a measure of fulfillment of the athletes' conditioning plans.

In our knowledge, the proposed methodology for the first time was applied for sports management process optimization. However, the scope of the methodology is not limited to the field of athletic training analysis and optimization. Another challenge in this study is the limited amount of data available. The data of two athletes' Olympic training cycles are available. In addition, the data are not independent, as both athletes compete in a pair. This complicates the application of other methods, such as statistical methods, to address this problem. Therefore, in this application, the visualization approach should be defined for monitoring the condition of top kayakers during the three-year macro cycles before the Olympic Games.

Except for the Introduction, the article is organized as follows. In Section 2, the description of the experiment and the measured and expected target values of the indicators set by experts are provided. Section 3 is focused on justifying the selection of the approximation function. In Section 4, the research methodology, based on the approximation of functions by using a linear combination of polynomials and trigonometric functions, is described, and the calculated unknown coefficients are given. Section 5 describes visualization of the approximation of the target and the measured functions, and presents a measure of the difference between the target and the measurement functions, which defines the athlete's condition. Section 6 provides the discussion of the investigated problem and conclusions.

2. The Investigation Object and the Initial Data

2.1. The athletes' Participation in the Study

The research was performed in the second, third, and the fourth year of the Olympic four-year cycle, at the time of athletic training for Rio de Janeiro Olympic Games in 2016. Two (1, 2) elite flat-water kayak paddlers (a racing team), performing at the international competitive level at the distance of 1000 m (whose ages were 27 and 26 years, and body mass were 88.5 and 84.5 kg, respectively), volunteered to take part in the investigation. These two athletes gained fifth place in a 1000 m event (K-2) in the 2016 Olympic Games in Rio de Janeiro.

2.2. The Description of the Experiment

A council, consisting of 22 Lithuanian kayaking elite experts (trainers and sports researchers), with no less than 10 years of experience in execution and organization in kayaking, have rated the competences and created the dynamics of the ideal curves of indicators based on correlations between indicators and sports outcomes [20].

The training volume and intensity were carefully controlled and quantified by using the Garmin Connect Forerunner 910 XT during each training session throughout the cycle. The data obtained from the devices were sent to three delegates from the sports science experts' council, who are the creators of the athletes' training programs. During the study, the athletes were encouraged to undertake their standard training sessions but not to train on the day before each test. The athletes were acquainted with the experimental procedures prior to testing and gave a written consent to participate in the study. All experimental procedures were approved by the Lithuanian Ethics Committee.

The testing lasted for three seasons (macro cycles): in each season, the training period lasted for 8 months, the competition period lasted for 3 months, and the transition period lasted for 1 month. Physiological testing of the athletes' condition was performed at the Lithuanian University of Education Science. For each athlete, all tests were conducted at the same time of the day, between 09:00 and 11:00, and 24 h after the last training session. During the testing session, the tests were performed in the same order. Prior to beginning the study, the sports doctor examined the kayakers to exclude any medical disorders that could limit their participation in the investigation.

During the macro cycles (I, II, III), the athletes were tested in six periods of time: in (T1), in the first week of the introductory training period, in (T2), at the beginning of the general training period, in (T3), at the beginning of the specific training period, in (T4), at the beginning of the competitive training period, in (T5), at the beginning of the main competition period, and in (T6), at the time after the main competition period.

Standard methodologies were applied to determine the athletes' physical parameters, i.e., their height (cm) and body mass (kg) [21].

The resting heart rate (b/min) was determined for each athlete in the supine position and in the period after the application of the standard physical load (30 squats within 45 s) by using the Garmin Connect Forerunner 910 XT.

Hemoglobin concentration (g/L) was determined by using a Hemocue analyzer. A trace amount of blood samples was taken from the fingertips of the athletes in the resting position.

The ergometer test was performed, using Oxycon Mobile 781023-052, version 5.2 (Cardinal Health Germany 234 GmbH, Höchberg, Germany). Gas analyzers were calibrated before and verified after each test. Each athlete performed the incremental submaximal ergometer test, using a calibrated kayak ergometer (Dansprint PRO, KE001 ergo, Hvidovre, Denmark) for determining the maximum oxygen uptake (VO_2max) [22]. Five minutes before performing the submaximal ergometer test, the athletes completed a 15-min warm up. The incremental test began with the application of the initial workload of 100 W, and increments of 20 W were applied at the intervals of 30 s to bring the athlete to the limit of tolerance in 8–12 min. During the last minute, the athlete was encouraged to do as much work as he could. Then, the values were averaged over the intervals of 30 s. Pulmonary ventilation (PV) (1/min), oxygen uptake (VO_2) (1/min, mL/min/kg), work capacity (W), and speed (km/h) were recorded at the point of the critical intensity limit (CIL).

For reaching the aim of the present study, all tests were arranged in hierarchical order by the experts' board (from the most informative to the least informative indicator), which influenced the outcome of a sporting event, measured in percent (Table 1). In addition, all the testing sessions were also arranged in the hierarchical order. Their importance is shown in Table 2.

Table 1. Test indicators presented in the hierarchical order set by the experts from the most to the least important indicator.

Hierarchic Value	Test Indicator	Importance in Percent
1	Speed (km/h) at the point of the CIL	30
2	Work capacity (W), at the point of the CIL	20
3	Oxygen uptake (1/min, mL/min/kg) at the point of the CIL	15
4	Pulmonary ventilation (1/min) at the point of the CIL	10
5	Resting heart rate (b/min)	10
6	Heart rate (b/min) after the standard physical load	10
7	Haemoglobin concentration (g/L)	5

Table 2. Importance of the testing sessions presented in the chronological order from the least (1) to the most (10) important session and their hierarchical order set by the experts.

No	Date	Testing Sessions	Hierarchical Order	Athletic Training Phase
1	2013 09 13	IT1	1	Transitional period after the season
2	2013 12 18	IT2	3	The first half of the training period (without water)
3	2014 03 28	IT3	4	The second half of the training period (with water)
4	2014 05 15	IT4	5	The period of non-core competitions
5	2014 07 15	IT5	6	European Championship
6	2014 08 12	IT6	7	World Championship
7	2014 09 12	IIT1	2	Transitional period after the season
8	2014 12 18	IIT2	4	The first half of the training period (without water)
9	2015 03 28	IIT3	5	The second half of the training period (with water)
10	2015 05 13	IIT4	6	The period of non-core competitions
11	2015 07 15	IIT5	7	European Championship
12	2015 08 12	IIT6	8	World Championship
13	2015 09 29	IIIT1	4	Transitional period after the season
14	2015 12 22	IIIT2	6	The first half of the training period (without water)
15	2016 03 04	IIIT3	7	The second half of the training period (with water)
16	2016 04 04	IIIT4	8	The period of non-core competitions
17	2016 07 07	IIIT5	9	European Championship/ Olympic selection
18	2016 08 01	IIIT6	10	Olympic Games

In Table 3, the results of measurements of seven indicators for both athletes $R_i^{(l,j)}$ and the expected target values $S_i^{(l,j)}$ determined by the experts are presented. Here, l denotes the athletes ($l = 1, 2$), j is the number of the indicator ($j = 1, 2, \ldots, 7$), and i is the number of the measurement ($i = 1, 2, \ldots, 18$). For example, $R_4^{(2,6)} = 105$, $R_2^{(1,3)} = 60.7$, $S_6^{(1,7)} = S_6^{(2,7)} = 168$. The expected target values $S_i^{(l,j)}$ differ for athletes 1 and 2 only in the case of $j = 4$ because of the different lung volumes of the athletes.

Table 3. Measurements of seven indicators for two athletes $R_i^{(l,j)}$ (the first row) and the expected target values $S_i^{(l,j)}$ (the second row).

i	j = 1		j = 2		j = 3		j = 4		j = 5		j = 6		j = 7	
1	260	260	180	185	60.9	48	186	146	48	56	106	126	158	180
	310		180		53		170	140	56		131		140	
2	280	280	195	185	60.7	48	189	149	48	48	109	132	164	179
	315		188		55		175	147	53.5		128		145	
3	290	290	200	200	61.8	44	198	167	44	52	106	130	167	163
	323		197		58		180	155	51		124		152	
4	300	300	205	200	64	56	187	167	56	56	111	105	146	170
	328		203		60		183	160	49		122		157	
5	320	320	213	205	62	48	189	171	48	52	109	129	165	173
	336		211		63		187	166	47		118		164	
6	320	320	215	215	57.2	48	186	170	48	48	119	118	134	162
	340		215		65		190	170	46		115		168	

Table 3. Cont.

i	j = 1		j = 2		j = 3		j = 4		j = 5		j = 6		j = 7	
7	300	300	190	212	62.1	48	159	159	48	60	111	114	156	169
	320		190		58		180	155	54		128		154	
8	310	310	195	195	59	48	195	156	48	52	114	120	159	177
	326		197		60.5		183	160	51.5		123		158	
9	320	320	210	211	55	56	188	163	56	56	119	128	161	175
	334		205		63.5		188	165	48.5		119		163	
10	340	340	216	216	63.7	44	193	160	44	52	113	116	166	174
	338		210		65		190	168	47		116		167	
11	320	320	220	220	62.6	48	193	155	48	44	113	125	156	172
	346		217		67		193	173	45		112		171	
12	340	340	220	221	63.9	48	191	162	48	52	110	121	163	182
	350		220		68		195	175	44		109		174	
13	300	300	200	195	60.2	52	193	161	52	48	122	117	174	178
	330		195		60		185	160	52		123		160	
14	320	320	210	205	54.1	44	192	167	44	56	103	120	166	176
	338		203		62.5		188.5	165	49.5		118		165	
15	340	340	217	216	50	48	185	164	48	52	108	108	159	164
	345		210		65		192	170	47		113		169	
16	350	350	220	221	59.4	48	190	172	48	48	111	112	167	162
	348		213		66		194	172	46		110		171	
17	340	340	224	224	58.2	52	192	157	52	44	113	106	164	180
	357		223		69		198.5	178	43		103		178	
18	340	340	217	217	57.8	52	181	154	52	48	121	102	160	167
	360		225		70		200	180	42		100		180	

3. The Selection of the Approximating Functions

The shape of the approximating functions was determined only for the target curves, which were created according to experts' judgments. The available data were compared with the ideal (target) curves. The following factors were taken into account for the selection of the suitable functions for the curves:

1. The approximation curve should not only describe the upward or downward trend, but also show the characteristic phase of the fall and the subsequent growth phase of the considered processes.
2. The shape of the functions should be the same for all the curves and be simple enough. Its parameter values were determined by the least squares method.

Let us show how the ideal curve was constructed for the case of RN2. The values of the normalized data (t_j, s_j), $j = 1, 2, \ldots, 18$ were as follows:

(0.0,1.0), (0.091,1.044), (0.186,1.094), (0.232,1.172), (0.290,1.150), (0.316,1.194), (0.346,1.056), (0.438,1.094), (0.533,1.139), (0.576,1.167), (0.636,1.206), (0.663,1.222), (0.708,1.083), (0.788,1.128), (0.857,1.167), (0.887,1.183), (0.976,1.239), (1.0,1.250).

The shape of the selected curve with the coefficients determined by the least squares method was as follows:
$$w(t) = 1.019 + 0.245t - 0.010t^2 + 0.044 \sin(2\pi t) - 0.013 \cos(2\pi t)$$

The curve $w(t)$ and the linear interpolation I of the points (t_j, s_j) are depicted in Figure 1. Note that the graphical accuracy is sufficient for the purpose of the study, because the curve $w(t)$ satisfies requirements 1 and 2. On the other hand, describing all the fluctuations of the linear interpolation I does not make sense, because the similarity of these real data points can be hardly realized in practice. It should also be noted that the authors did not intend to construct the best curves, and therefore constructed the curves of sufficient approximation. However, other combinations of algebraic and trigonometric polynomials, which were acceptable, were also tested. The final decision about the choice of the formula was taken according to requirement 2 (the formula is appropriate for all the cases considered).

Figure 2 shows the functions

$$y(t) = 1.133 + 0.020t, \; z(t) = 1.046 + 0.24183t - 0.081t^2$$

with the coefficients determined by the least squares method based on the same data. We can see that the curves $y(t)$ and $z(t)$ do not satisfy requirement 1. The values of the error function were calculated as follows:

$$S_f = \sqrt{\sum_{i=1}^{18} (s_j - f(t_j))^2},$$

where f is equal to $y(t), z(t), w(t)$. The following error function values were obtained: $S_y = 0.0642$, $S_z = 0.0487$, $S_w = 0.0430$. One can see that the best result was achieved by incorporating trigonometric components into the approximation function.

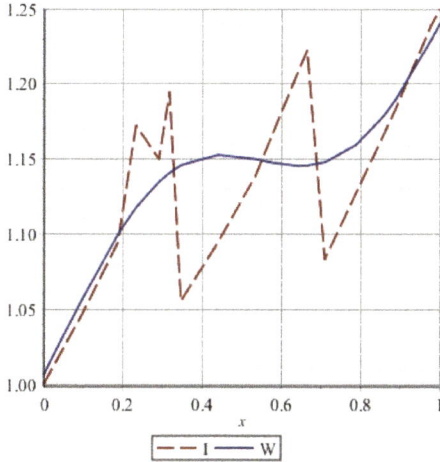

Figure 1. Approximation $w(t)$ of the target function I.

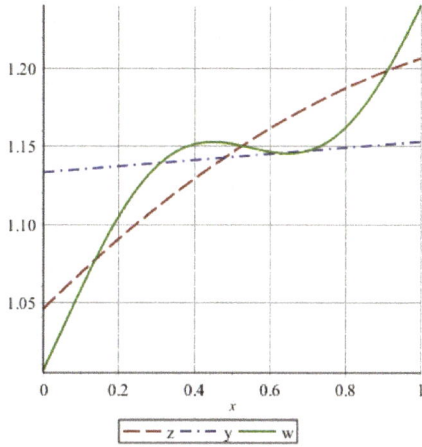

Figure 2. Approximations $y(t)$, $z(t)$, $w(t)$ of the target function I.

4. Research Methodology

In the first step, all the initial data points (Table 3) were normalized to eliminate the influence of the measurement units on the final results. The values $R_i^{(l,j)}$ and $S_i^{(l,j)}$ were normalized by the equations:

$$\tilde{r}_i^{(l,j)} = \frac{R_i^{(l,j)}}{S_1^{(l,j)}}, \quad \tilde{s}_i^{(l,j)} = \frac{S_i^{(l,j)}}{S_1^{(l,j)}}$$

This method of data normalization, when each measurement $R_i^{(l,j)}$ and target value $S_i^{(l,j)}$ were divided by the first target value of the respective measurement, was chosen for the convenience of data representation. In this case, the graphs of all the target values of the indicators have the same initial point of 1.

The measurement time was normalized as follows:

$$t_i = \frac{T_i}{T_{18}}$$

where T_i is the number of days passed since the first measurement, $T_1 = 0$. The respective values are shown in Table 4.

Table 4. Normalized time values t_i.

T_i	0	96	196	244	305	333	364	461	561
$t_i = T_i/T_{18}$	0	0.0912	0.1861	0.2317	0.2896	0.3162	0.3457	0.4378	0.5328
T_i	607	670	698	746	830	903	934	1028	1053
$t_i = T_i/T_{18}$	0.5764	0.6363	0.6629	0.7085	0.7882	0.8576	0.8870	0.9763	1.000

The goal was to find the functions most suitable for approximating the athlete's indicators. The target values for a multiple measure of conditioning were compared to the targets set by the experts. The regression analysis was used to model how well their conditioning could fit the target. The weighted least square metric is a measure of the fulfillment of the athletes' conditioning plans. The expression with five undefined coefficients, a, b, c, A, B, was chosen for approximating the functions $\tilde{r}^{(l,j)}(t)$, $\tilde{s}^{(l,j)}(t)$ as follows:

$$f(t) = a + bt + ct^2 + A\sin 2\pi t + B\cos 2\pi t \quad (1)$$

where $t \in [0; 1]$ is the normalized time. The unknown coefficients were determined by a standard method of the least squares [23]:

$$\sum_{i=1}^{18}(f(t_i) - f_i)^2 \to min. \quad (2)$$

The system of linear equations was obtained from Equations (1) and (2) as follows:

$$\begin{pmatrix} a_{11} & a_{12} & a_{13} & a_{14} & a_{15} \\ a_{21} & a_{22} & a_{23} & a_{24} & a_{25} \\ a_{31} & a_{32} & a_{33} & a_{34} & a_{35} \\ a_{41} & a_{42} & a_{43} & a_{44} & a_{45} \\ a_{51} & a_{52} & a_{53} & a_{54} & a_{55} \end{pmatrix} \cdot \begin{pmatrix} a \\ b \\ c \\ A \\ B \end{pmatrix} = \begin{pmatrix} b_1 \\ b_2 \\ b_3 \\ b_4 \\ b_5 \end{pmatrix}$$

where

$$a_{11} = \sum_{i=1}^{18} 1 = 18, \quad a_{12} = a_{21} = \sum_{i=1}^{18} t_i, \quad a_{13} = a_{31} = \sum_{i=1}^{18} t_i^2, \quad a_{14} = a_{41} = \sum_{i=1}^{18} \sin 2\pi t_i, \quad a_{15} = a_{51} = \sum_{i=1}^{18} \cos 2\pi t_i,$$

$$a_{22} = \sum_{i=1}^{18} t_i^2, \quad a_{23} = a_{32} = \sum_{i=1}^{18} t_i^3, \quad a_{24} = a_{42} = \sum_{i=1}^{18} t_i \sin 2\pi t_i, \quad a_{25} = a_{52} = \sum_{i=1}^{18} t_i \cos 2\pi t_i, \quad a_{33} = \sum_{i=1}^{18} t_i^4,$$

$$a_{34} = a_{43} = \sum_{i=1}^{18} t_i^2 \sin 2\pi t_i, \quad a_{35} = a_{53} = \sum_{i=1}^{18} t_i^2 \cos 2\pi t_i, \quad a_{44} = \sum_{i=1}^{18} \sin^2 2\pi t_i, \quad a_{45} = a_{54} = \sum_{i=1}^{18} \sin 2\pi t_i \times \cos 2\pi t_i, \quad a_{55} = \sum_{i=1}^{18} \cos^2 2\pi t_i.$$

The coefficients $b_1, b_2, \ldots, b_5$ were calculated by the equations:

$$b_1 = \sum_{i=1}^{18} f_i, \quad b_2 = \sum_{i=1}^{18} f_i t_i, \quad b_3 = \sum_{i=1}^{18} f_i t_i^2, \quad b_4 = \sum_{i=1}^{18} f_i \sin 2\pi t_i, \quad b_5 = \sum_{i=1}^{18} f_i \cos 2\pi t_i,$$

where the functions f_i obtained the values $\widetilde{r}^{(l,j)}(t)$ and $\widetilde{s}^{(l,j)}(t)$, i.e., the coefficients a, b, c, A, B were calculated separately for each athlete $l = 1, 2$ and each indicator $j = 1, 2, \ldots, 7$ for the functions $\widetilde{r}$ and $\widetilde{s}$.

The matrix $A = (a_{ij})_{5 \times 5}$, depending only on the values t_i, was calculated only once:

$$A = \begin{pmatrix} 18 & 9.5242 & 6.6505 & -0.2485 & 0.5155 \\ 9.5242 & 6.6505 & 5.1966 & -2.7916 & 0.7607 \\ 6.6505 & 5.1966 & 4.3262 & -2.7751 & 1.6716 \\ -0.2485 & -2.7916 & -2.7751 & 9.2166 & -0.0602 \\ 0.5155 & 0.7607 & 1.6716 & -0.0602 & 8.7834 \end{pmatrix}$$

For calculating the coefficients a, b, c, A, B, the inverse matrix method is convenient to use (see, for example, [24]):

$$\begin{pmatrix} a \\ b \\ c \\ A \\ B \end{pmatrix} = A^{-1} \cdot \begin{pmatrix} b_1 \\ b_2 \\ b_3 \\ b_4 \\ b_5 \end{pmatrix} \tag{3}$$

$$A^{-1} = \begin{pmatrix} 2.7001 & -14.5727 & 13.8234 & -0.1888 & -1.5284 \\ -14.5727 & 85.2846 & -83.4043 & 0.3864 & 9.3445 \\ 13.8234 & -83.4043 & 82.7523 & -0.0333 & -9.3368 \\ -0.1888 & 0.3864 & -0.0333 & 0.2103 & -0.0146 \\ -1.5284 & 9.3445 & -9.3368 & -0.0146 & 1.7111 \end{pmatrix}$$

In Table 5, the values of the coefficients a, b, c, A, B, which were calculated by Equation (3), are presented for $j = 1, 2, \ldots, 7$, $l = 1, 2$ for the functions $\widetilde{r}$ and $\widetilde{s}$.

Table 5. Calculated values of the coefficients a, b, c, A, B.

			a	b	c	A	B
$j = 1$	$l = 1$	$\widetilde{r}$	0.835	0.529	-0.257	0.030	0.002
		$\widetilde{s}$	1.011	0.094	0.062	0.018	-0.010
	$l = 2$	$\widetilde{r}$	0.915	0.169	0.019	0.028	-0.007
		$\widetilde{s}$	1.011	0.094	0.062	0.018	-0.010
$j = 2$	$l = 1$	$\widetilde{r}$	0.976	0.651	-0.446	0.022	0.036
		$\widetilde{s}$	1.019	0.245	-0.010	0.044	-0.013
	$l = 2$	$\widetilde{r}$	1.038	0.237	-0.018	0.032	-0.023
		$\widetilde{s}$	1.019	0.245	-0.010	0.044	-0.013
$j = 3$	$l = 1$	$\widetilde{r}$	1.207	-0.323	0.240	0.007	-0.049
		$\widetilde{s}$	1.062	0.045	0.267	0.041	-0.058
	$l = 2$	$\widetilde{r}$	1.024	0.795	-0.836	0.033	0.052
		$\widetilde{s}$	1.062	0.045	0.267	0.041	-0.058

Table 5. Cont.

			a	b	c	A	B
$j = 4$	$l = 1$	$\widetilde{r}$	1.079	0.244	−0.279	−0.030	0.033
		$\widetilde{s}$	1.017	0.140	0.031	0.021	−0.014
	$l = 2$	$\widetilde{r}$	0.876	1.467	−1.368	0.033	0.133
		$\widetilde{s}$	1.023	0.312	−0.035	0.041	−0.021
$j = 5$	$l = 1$	$\widetilde{r}$	0.930	−0.501	0.569	0.029	−0.067
		$\widetilde{s}$	0.969	−0.145	−0.080	−0.037	0.018
	$l = 2$	$\widetilde{r}$	0.952	0.029	−0.136	−0.005	−0.005
		$\widetilde{s}$	0.969	−0.145	−0.080	−0.037	0.018
$j = 6$	$l = 1$	$\widetilde{r}$	0.907	−0.484	0.561	0.021	−0.085
		$\widetilde{s}$	0.962	0.089	−0.310	−0.030	0.031
	$l = 2$	$\widetilde{r}$	0.990	−0.058	−0.154	−0.047	0.001
		$\widetilde{s}$	0.962	0.089	−0.310	−0.030	0.031
$j = 7$	$l = 1$	$\widetilde{r}$	1.125	0.151	−0.167	−0.047	0.032
		$\widetilde{s}$	1.029	0.214	0.066	0.036	−0.029
	$l = 2$	$\widetilde{r}$	1.389	−0.779	0.692	−0.042	−0.081
		$\widetilde{s}$	1.029	0.214	0.066	0.036	−0.029

5. Visualization of the Results and Measurement of the Athlete's Condition

The graphs of approximations of the functions $\widetilde{r}(t)$ and $\widetilde{s}(t)$, with the coefficient values a, b, c, A, B, which are given in Table 5, are presented in Figures 3–9. The graphs of the approximations of the function $\widetilde{r}(t)$ are depicted in black, while the approximations of the function $\widetilde{s}(t)$ are shown in green.

Visualization of the results provides information about the quality of the athletes' training management and gives an understanding of the time periods when the maximum discrepancy between the ideal and the real curves can be observed. This allows the coaches to quickly change the athletes' workouts. To assess the numerical value of this mismatch, the measurement of the difference between the respective functions is required.

It is worth noting that five indicators have a direct relationship with the target, while two indicators, i.e., resting heart rate (5) and heart rate after the standard physical load (6), have the inverse relation with it. By analyzing speed at the point of the CIL (Figure 3), it could be observed that the athlete RN could consistently approach the target (ideal) indicators, although the AO progress was much slower. The tested athletes could not maximally synchronize their efforts (with respect to the above indicator). This can be accounted for by the specific character of their adaptation.

The evaluation of the athletes' work capacity at the point of the CIL (Figure 4) allows the authors to state that the dynamics of this indicator for the tested athletes was close to the target values, and they achieved their planned peak performance before the beginning of the Olympic Games.

Evaluation of the dynamics of the criteria describing oxygen uptake at the point of the CIL (Figure 5) showed the decreasing trend of VO_2max during the testing period and its considerable difference from the target value. This can be accounted for by the specific character of the athletes' training aimed at increasing the load in the muscles' area where glycolytic energy is generated. This phenomenon requires further investigation.

When evaluating the variation in the criteria values of pulmonary ventilation at the point of the CIL (Figure 6), a similar trend could be observed as that characteristic of the criteria describing oxygen uptake at the point of the CIL. It can be assumed that the athletes' aerobic capacity had a tendency to decrease. This can be associated with the striving of athletes to increase their capacity of energy generation in the training period.

Evaluation of the variation of criteria describing the resting heart rate (Figure 7) in a resting position and the heart rate after the application of a standard physical load (Figure 8) showed that the adaptation of the athletes' circulatory system to training was different. Thus, it was unstable for RN, while AO consistently approached the ideal values of indicators.

The analysis of the variation in hemoglobin concentration of the athletes (Figure 9) has shown that this indicator was difficult to control. In the training process, hemoglobin concentration varied considerably. As a result, the target values could not be achieved before the main competitions.

In the ideal case, when the athlete's training plan is fully realized, the curves $\widetilde{r}(t)$ and $\widetilde{s}(t)$ coincide. In practice, this objective can only be achieved partially; therefore, it is necessary to quantify the difference between the real and the ideal curves.

This difference is a measure of the athletes' condition. The fact that the differences $|\widetilde{r}(t) - \widetilde{s}(t)|$ observed in various time intervals are of various importance is taken into account. The relative importance of the time intervals was determined in the following way. The relative importance of the period prior to measuring was equal to 1, while the importance of other periods was higher, ranging from 2 to 10. The definitions of the time intervals, their relative importance, and the dates and lengths of the intervals (in percent) are given in Table 6.

Table 6. Relative importance of the time intervals.

Date	Definition of the Time Interval	Relative Importance of the Interval	Interval	The Interval's Length, in Percent
2013 09 13	The transitional period after the season	1	Till T_1	-
2013 12 18	The first half of the training period (without water)	3	$[T_1, T_2]$	9.1
2014 03 28	The second half of the training period (with water)	4	$[T_2, T_3]$	9.5
2014 05 15	The period of non-core competitions	5	$[T_3, T_4]$	4.6
2014 07 15	European Championship	6	$[T_4, T_5]$	5.8
2014 08 12	World Championship	7	$[T_5, T_6]$	2.7
2014 09 12	The transitional period after the season	2	$[T_6, T_7]$	2.9
2014 12 18	The first half of the training period (without water)	4	$[T_7, T_8]$	9.2
2015 03 28	The second half of the training period (with water)	5	$[T_8, T_9]$	9.5
2015 05 13	The period of non-core competitions	6	$[T_9, T_{10}]$	4.4
2015 07 15	European Championship	7	$[T_{10}, T_{11}]$	6.0
2015 08 12	World Championship	8	$[T_{11}, T_{12}]$	2.7
2015 09 29	The transitional period after the season	4	$[T_{12}, T_{13}]$	4.6
2015 12 22	The first half of the training period (without water)	6	$[T_{13}, T_{14}]$	7.9
2016 03 04	The second half of the training period (with water)	7	$[T_{14}, T_{15}]$	6.9
2016 04 04	The period of non-core competitions	8	$[T_{15}, T_{16}]$	2.9
2016 07 07	European Championship/Olympic selection	9	$[T_{16}, T_{17}]$	8.9
2016 08 01	Olympic Games	10	$[T_{17}, T_{18}]$	2.4

Depending on the relative importance of the time intervals and their relative length, the weights w_k were assigned to the normalized time intervals $[t_{k-1}, t_k]$. In this case, $t_0 = 0$, $t_k = \frac{k}{100}$ and $k = 1, 2, \ldots, 100$. The weights of the time intervals are given in Table 7.

Table 7. The weights of the time intervals.

w_1	w_2	$w_3 = \ldots = w_9$	$w_{10} = \ldots = w_{18}$	$w_{19} = \ldots = w_{23}$	$w_{24} = \ldots = w_{29}$	$w_{30} = \ldots = w_{32}$	w_{33}
0.01	0.02	0.03	0.04	0.05	0.06	0.07	0.06
w_{34}	w_{35}	w_{36}	w_{37}	w_{38}	$w_{39} = \ldots = w_{44}$	$w_{45} = \ldots = w_{53}$	$w_{54} = \ldots = w_{57}$
0.05	0.04	0.03	0.02	0.03	0.04	0.05	0.06
$w_{58} = \ldots = w_{63}$	$w_{64} = \ldots = w_{66}$	w_{67}	w_{68}	w_{69}	$w_{70} = \ldots = w_{71}$	w_{72}	$w_{73} = \ldots = w_{79}$
0.07	0.08	0.07	0.06	0.05	0.04	0.05	0.06
$w_{80} = \ldots = w_{86}$	$w_{87} = \ldots = w_{89}$	$w_{90} = \ldots = w_{98}$	$w_{99} = \ldots = w_{100}$				
0.07	0.08	0.09	0.10				

It should be emphasized that the fluctuations of the relative importance values in Table 6 were smoothed out. For example, $w_{66} = 0.08$ and the intermediate values $w_{67} = 0.07$, $w_{68} = 0.06$, $w_{69} = 0.05$ appeared before $w_{70} = 0.04$, since it was difficult to detect the exact moment of their sudden decrease.

Thus, the difference between the functions $\widetilde{r}(t)$ and $\widetilde{s}(t)$ was measured as the weighted sum as follows:

$$\|\widetilde{r}(t) - \widetilde{s}(t)\| = \frac{\sum_{k=1}^{100} w_k |\widetilde{r}(t_k) - \widetilde{s}(t_k)|}{\sum_{k=1}^{100} w_k \widetilde{s}(t_k)}.$$

The values $\|\widetilde{r}(t) - \widetilde{s}(t)\|$ calculated for each indicator and both athletes are given in Table 8.

Table 8. The values $\|\widetilde{r}(t) - \widetilde{s}(t)\|$ for seven indicators and two athletes.

j	$l = 1$	$l = 2$
1	0.066	0.053
2	0.011	0.015
3	0.085	0.084
4	0.046	0.024
5	0.067	0.049
6	0.019	0.062
7	0.058	0.039

The results provided in Table 8 are summarized as the weighted sums $S^{(l)}$, $l = 1, 2$, with the respective criteria $j = 1, 2, \ldots, 7$ weights: $\widetilde{w}_1 = 0.3$, $\widetilde{w}_2 = 0.2$, $\widetilde{w}_3 = 0.15$, $\widetilde{w}_4 = \widetilde{w}_5 = \widetilde{w}_6 = 0.1$, $\widetilde{w}_7 = 0.05$ (see the data in Table 1). The weighted sums were as follows:

$$S^{(l)} = \sum_{j=1}^{7} \widetilde{w}_j \|\widetilde{r}^{(l,j)}(t) - \widetilde{s}^{(l,j)}(t)\|. \tag{4}$$

Therefore, $S^{(1)} = 0.051$, $S^{(2)} = 0.047$. The values $S^{(l)}$ have an inverse relationship with the athletic training. The higher these values, the lower the achievement of the goals by the athlete. Then, the values of $K^{(l)} = 1 - S^{(l)}$, describing the results of the athletic training, were calculated as follows:

$$K^{(1)} = 0.949, \quad K^{(2)} = 0.953.$$

The higher the value of the indicator K, the higher the level of the fulfillment of the athlete's training plan. It can be concluded that the results of the second athlete are better than the results obtained by the first athlete. The indicator K is important for the implementation and improvement (optimization) of the athlete's training plan development.

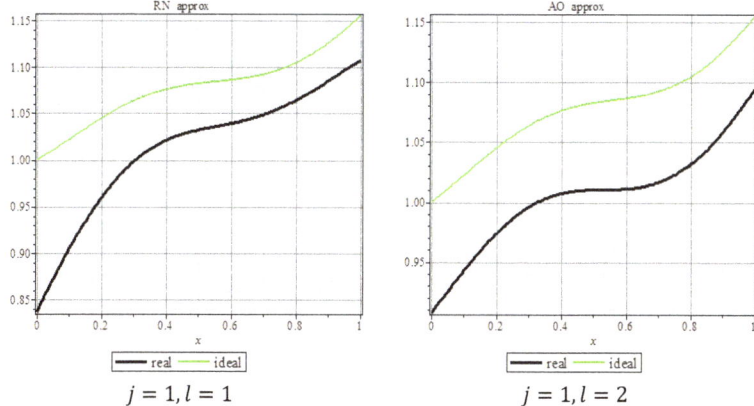

Figure 3. Approximation of functions $\widetilde{r}(t)$ and $\widetilde{s}(t)$ for two athletes according to speed (km/h) at the point of the CIL.

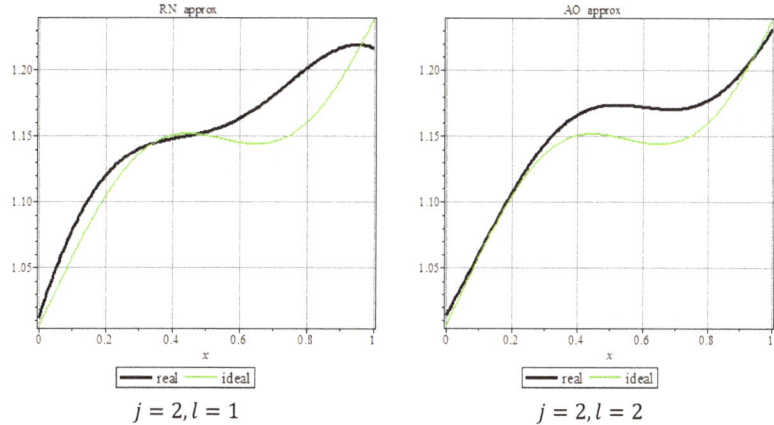

Figure 4. Approximation of functions $\widetilde{r}(t)$ and $\widetilde{s}(t)$ for two athletes based on their work capacity (W) at the point of the CIL.

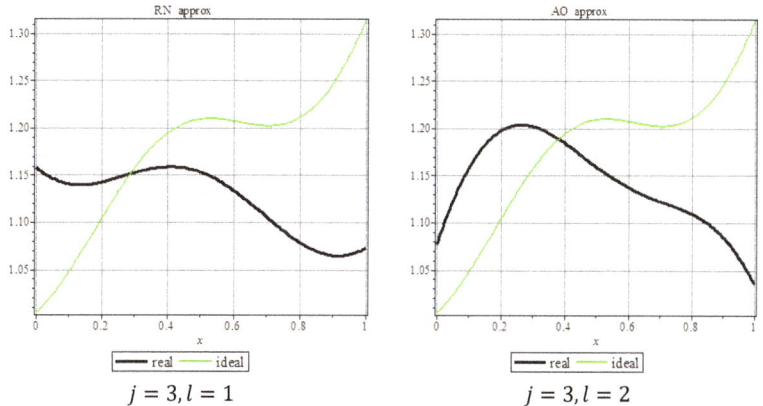

Figure 5. Approximation of functions $\widetilde{r}(t)$ and $\widetilde{s}(t)$ for two athletes based on oxygen uptake (1/min, mL/min/kg) at the point of the CIL.

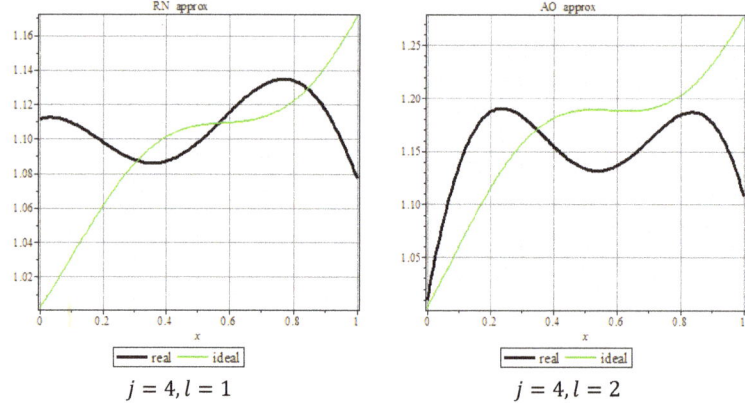

Figure 6. Approximation of functions $\widetilde{r}(t)$ and $\widetilde{s}(t)$ for two athletes based on pulmonary ventilation (1/min) at the point of the CIL.

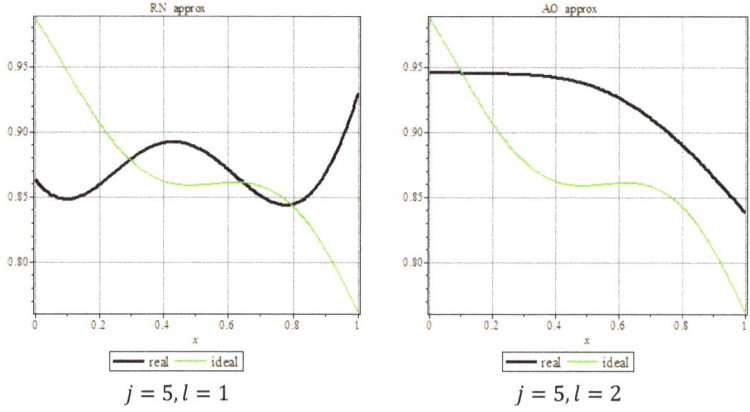

Figure 7. Approximation of the functions $\widetilde{r}(t)$ and $\widetilde{s}(t)$ for two athletes based on the resting heart rate (b/min).

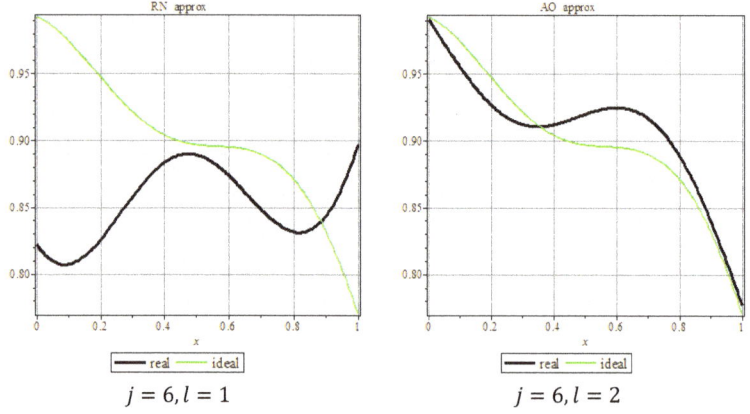

Figure 8. Approximation of the functions $\widetilde{r}(t)$ and $\widetilde{s}(t)$ for two athletes based on their heart rate (b/min) after the standard physical load.

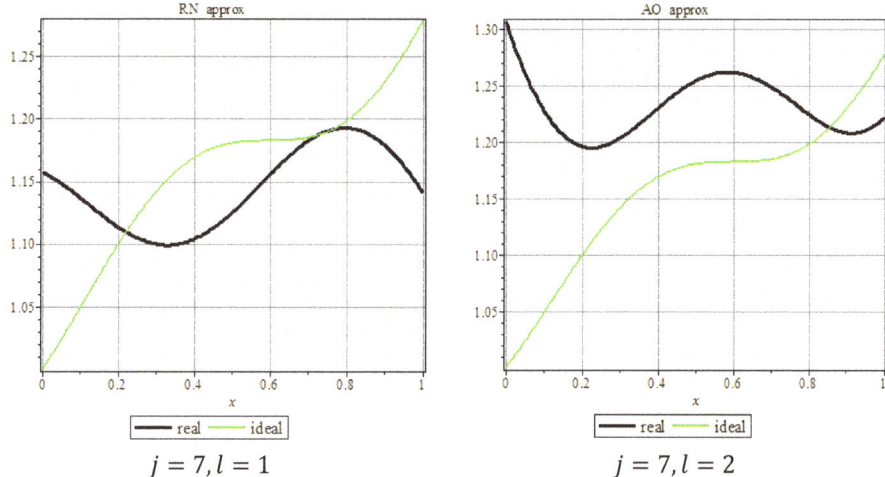

Figure 9. Approximation of functions $\widetilde{r}(t)$ and $\widetilde{s}(t)$ for two athletes based on haemoglobin concentration (g/L).

6. Discussion and Conclusions

The optimization and evaluation of athletic training has become an increasingly important problem for sports scientists, coaches, and athletes [25]. The character of the identification factors of the elite athlete's condition and their integration into the processes of strategic planning, management, and parameter estimation have not been thoroughly investigated, but they are obviously important. To govern the process of elite athletes' training, it is required to consider this process both as a whole and in detail. Not all the aspects of the executed cycle can be in compliance. Therefore, the research should not be performed in terms of the standardized compliance management. The OR methods allow us to define the criteria and the limits of their permissible pursuit [14].

The papers considered do not offer a unique view of past research, and the integration they provide is likely to be only fractional. Therefore, real research data are needed. Usually, the research data seem to be more significant than dimensions. They present (1) the constructs and their relational properties and (2) consistency in the measurement of phenomena. The combinations of these dimensions can yield the following results: (1) theoretical plurality and empirical convergence, (2) the dominant paradigm, (3) fragmentation, and (4) the convergent theory and empirical plurality. The integration will require fostering both taxonomic and methodological commensurability among the different sub-fields and contributing disciplines [26]. The method of parameter estimation of the dynamics of the factors presents a viable research approach in this context [27]. It is particularly useful for the investigation of the process cycle, e.g., its long-term effects and results [28]. Interactive data visualization is an evolving approach, providing wide opportunities for managing and decision making in the case of multidimensional decision problems and planning processes.

The proposed methodology of the athletic training visualization makes it easier to identify the advantages and disadvantages of training and shows when (at what stages of training) physical loads (or training methods and techniques) do not match the adaptive capacities of the tested athletes. It also provides new opportunities for streamlining and optimizing the key factors influencing the process by modifying its components. The authors believe that data visualization allows for achieving a better way of informing athletes and coaches about the mistakes and the right choices in the considered training period. According to Xu and Ding [29], visualization allows for identifying the deviations of the indicator values. The automated alarm system, warning about critical deviations of the process components, could help to optimize the process management and prevent hazardous incidents [30].

To the author's best knowledge, no cases of applying the methodology based on the goal function's approximation by a linear combination of polynomials and trigonometric functions can be found in the scientific literature on sports management. Meanwhile, this methodology, in conjunction with the least squares method, yields fairly accurate approximation results.

The described methodology (algorithm) has a number of advantages over other similar approaches. The visualization of measurement and target approximation functions enables sports scientists and coaches to perform the following functions:

- to supervise the athletes' training process,
- to identify undesirable tendencies,
- to correct them immediately (if necessary).

The proposed Formula (4) for measuring the differences between the indicators and the current and target functions enables the researchers to identify the athlete's condition, as well as showing his/her potential and providing more accurate predictions of sports results.

New opportunities are provided by the proposed method, which:

- Facilitate the understanding and achievement of the goal;
- Make processes clearer to the general public;
- Allow for effective management of the processes.

The provided high-quality, effective alternative solutions to the problem of achieving the final goals of the considered management process give a deeper insight into the effectiveness of a solution, which can support a decision-making process. The principles of the model of the parameter estimation provide visualization capabilities, allowing for evaluating the possibilities to reach the goal, which is the effective management process. This method also allows for predicting the future achievements of athletes.

The proposed model for evaluating the planning of athletic training is used in other sports and can be applied not only to sports practice, but also to the implementation of many various strategies. Moreover, based on the proposed visualization method, an application could be created that would make this method a more widely used technique.

Author Contributions: Conceptualization, S.D.; Data curation, A.K., S.D. and N.K.; Formal analysis, R.D.; Methodology, A.K.; Supervision, S.D.; Visualization, A.K.; Writing—original draft, N.K., R.D.; Writing—review & editing, N.K. and S.D. All authors have read and agreed to the published version of the manuscript.

Funding: This research received no external fundingNo external funding.

Conflicts of Interest: The authors declare no conflict of interest.

References

1. Issurin, V.B. Evidence-based prerequisites and precursors of athletic talent: A review. *Sports Med.* **2017**, *47*, 1993–2010. [CrossRef] [PubMed]
2. Jurimae, J. Methods for monitoring training status and their effects on performance in rowing. *Int. SportMed J.* **2008**, *9*, 11–21.
3. Latyshev, M.; Latyshev, S.; Kvasnytsya, O.; Knyazev, A. Performance analysis of freestyle wrestling competitions of the last Olympic cycle. *J. Phys. Educ. Sport* **2017**, *17*, 590–594.
4. Guellich, A.; Seiler, S.; Emrich, E. Training methods and intensity distribution of young world-class rowers. *Int. J. Sports Physiol.* **2009**, *4*, 448–460.
5. Bompa, T.O.; Haff, G. *Periodization: Theory and Methodology of Training*, 6th ed.; Human Kinetics: Champaign, IL, USA, 2018.
6. Zatsiorsky, V.M.; Kraemer, W.J. *Science and Practice of Strength Training*; Human Kinetics: Champaign, IL, USA, 2006.
7. Halson, S.L. Monitoring training load to understand fatigue in athletes. *Sports Med.* **2014**, *44*, 139–147. [CrossRef]

8. Polak, E.; Kulasa, J.; Vencesbrito, A.; Castro, M.; Fernandes, O. Motion analysis systems as optimization training tools in combat sports and martial arts. *Rev. Artes Marciales Asiáticas* **2015**, *10*, 105–123. [CrossRef]
9. Fister, I.; Rauter, S.; Yang, X.-S.; Ljubic, K. Planning the sports training sessions with the bat algorithm. *Neurocomputing* **2015**, *149*, 993–1002. [CrossRef]
10. Dadelo, S.; Turskis, Z.; Zavadskas, E.K.; Kačerauskas, T.; Dadelienė, R. Is the Evaluation of the Students' Values Possible? An Integrated Approach to Determining the Weights of Students' Personal Goals Using Multiple-Criteria Methods. *Eurasia J. Math. Sci. Technol.* **2016**, *12*, 2771–2781. [CrossRef]
11. Armstrong, K.J.; Weidner, T.G.; Walker, S.E. Athletic training approved clinical instructors' reports of real-time opportunities for evaluating clinical proficiencies. *J. Athl. Train.* **2009**, *44*, 630–638. [CrossRef]
12. Li, F.; Zhu, Q.; Chen, Z.; Xue, H.A. A balanced data envelopment analysis cross-efficiency evaluation approach. *Expert Syst. Appl.* **2018**, *106*, 154–168. [CrossRef]
13. Karsu, Ö.; Morton, A. Incorporating balance concerns in resource allocation decisions: A bi-criteria modelling approach. *Omega* **2014**, *44*, 70–82. [CrossRef]
14. Reefke, H.; Sundaram, D. Key themes and research opportunities in sustainable supply chain management—Identification and evaluation. *Omega* **2017**, *66*, 195–211. [CrossRef]
15. Recalde, D.; Torres, R.; Vaca, P. Scheduling the professional Ecuadorian football league by integer programming. *Comput. Oper. Res.* **2013**, *40*, 2478–2484. [CrossRef]
16. Dadelo, S.; Turskis, Z.; Zavadskas, E.K.; Dadelienė, R. Multi-criteria assessment and ranking system of sport team formation based on objective-measured values of criteria set. *Expert. Syst. Appl.* **2014**, *41*, 6106–6113. [CrossRef]
17. Jones, C. Visualization and Optimization. *J. Oper. Res. Soc.* **1997**, *48*, 964. [CrossRef]
18. Laengle, S.; Merigó, M.M.; Miranda, J.; Słowiński, R.; Bomze, I.; Borgonovo, E.; Dyson, R.G.; Oliveira, J.F.; Teunter, R. Forty years of the European Journal of Operational Research: A bibliometric overview. *Eur. J. Oper. Res.* **2017**, *262*, 803–816. [CrossRef]
19. Moritz, D.; Fisher, D.; Ding, B.; Wang, C. Trust, but verify optimistic visualizations of approximate queries for exploring big data. In Proceedings of the 2017 CHI Conference on Human Factors in Computing Systems, CHI'17, Denver, CO, USA, 6–11 May 2017; pp. 2904–2915.
20. Balčiūnas, E. *Didelio Meistriškumo Baidarininkų Rengimas 200 ir 500 m Nuotoliams: Monografija/Training of Elite Kayakers for Distances of 200 and 500 m: The Monograph*; Lietuvos edukologijos universiteto leidykla: Lithuania, Vilnius, 2016. (In Lithuania)
21. Norton, K.; Olds, T. *Antropometrica*; University of New South Wales Press: Sydney, Australia, 1996.
22. Thodens, J.S. *Testing Aerobic Power. Physiological Testing of the High-Performance Athlete*; Human Kinetics: Champaign, IL, USA, 1991.
23. Legendre, A.-M. *Nouvelles Méthodes pour la Détermination des Orbites des Comètes [New Methods for the Determination of the Orbits of Comets]*; F. Didot: Paris, France, 1805. (In French)
24. Strang, G. *Linear Algebra and Its Applications*, 4th ed.; Cengage Learning: Boston, MA, USA, 2006.
25. Spotts, J.D. Global politics and the Olympic Games: Separating the two oldest games in history. *Dickinson J. Int. Law* **1994**, *13*, 103–122.
26. Durand, R.; Grant, R.M.; Madsen, T.L. The expanding domain of strategic management research and the quest for integration. *Strateg. Manag. J.* **2017**, *38*, 4–16. [CrossRef]
27. Krylovas, A.; Dadeliene, R.; Kosareva, N.; Dadelo, S. Comparative evaluation and ranking of the European countries based on the interdependence between human development and internal security indicators. *Mathematics* **2019**, *7*, 293. [CrossRef]
28. Sivarajah, U.; Kamal, M.M.; Irani, Z.; Weerakkody, V. Critical analysis of Big Data challenges and analytical methods. *J. Bus. Res.* **2017**, *70*, 263–286. [CrossRef]
29. Xu, L.; Ding, F. Recursive least squares and multi-innovation stochastic gradient parameter estimation methods for signal modelling. *Circuits Syst. Signal Process.* **2017**, *36*, 1735–1753. [CrossRef]
30. Krylovas, A.; Dadelo, S.; Kosareva, N.; Zavadskas, E.K. Entropy-KEMIRA approach for MCDM problem solution in human resources selection task. *Int. J. Inf. Technol. Decis. Mak.* **2017**, *16*, 1183–1209. [CrossRef]

© 2020 by the authors. Licensee MDPI, Basel, Switzerland. This article is an open access article distributed under the terms and conditions of the Creative Commons Attribution (CC BY) license (http://creativecommons.org/licenses/by/4.0/).

Article

Dynamic Modelling of Interactions between Microglia and Endogenous Neural Stem Cells in the Brain during a Stroke

Awatif Jahman Alqarni [1,2], Azmin Sham Rambely [1,*] and Ishak Hashim [1]

[1] Department of Mathematical Sciences, Faculty of Science and Technology, Universiti Kebangsaan Malaysia, 43600 UKM Bangi, Selangor, Malaysia; aw-aw910@hotmail.com (A.J.A.); ishak_h@ukm.edu.my (I.H.)
[2] Department of Mathematics, College of Sciences and Arts in Blqarn, University of Bisha, Bisha 61922, Saudi Arabia
* Correspondence: asr@ukm.edu.my; Tel.: +60-389213244

Received: 26 October 2019; Accepted: 11 December 2019; Published: 16 January 2020

Abstract: In this paper, we study the interactions between microglia and neural stem cells and the impact of these interactions on the brain cells during a stroke. Microglia cells, neural stem cells, the damage on brain cells from the stroke and the impacts these interactions have on living brain cells are considered in the design of mathematical models. The models consist of ordinary differential equations describing the effects of microglia on brain cells and the interactions between microglia and neural stem cells in the case of a stroke. Variables considered include: resident microglia, classically activated microglia, alternatively activated microglia, neural stem cells, tissue damage on cells in the brain, and the impacts these interactions have on living brain cells. The first model describes what happens in the brain at the stroke onset during the first three days without the generation of any neural stem cells. The second model studies the dynamic effect of microglia and neural stem cells on the brain cells following the generation of neural stem cells and potential recovery after this stage. We look at the stability and the instability of the models which are both studied analytically. The results show that the immune cells can help the brain by cleaning dead cells and stimulating the generation of neural stem cells; however, excessive activation may cause damage and affect the injured region. Microglia have beneficial and harmful functions after ischemic stroke. The microglia stimulate neural stem cells to generate new cells that substitute dead cells during the recovery stage but sometimes the endogenous neural stem cells are highly sensitive to inflammatory in the brain.

Keywords: chemokines; cytokines; eigenvalue stability analysis; neurogenesis; numerical solution; system of ordinary differential equations

---

1. Introduction

Strokes are diseases that affect the brain, which may cause death and disabilities in mammals [1–3]. When ischemic stroke occurs, the cells surrounding the ischemic area start to die once the outflow of blood declines below 10% of the normal blood influx [1,2]. Cells can die through necrosis or apoptosis [1,4]. Necrosis occurs very quickly after stroke onset; the necrotic cells lead to pollution of the local environment in the brain and damage the surrounding cells through laceration and the release of intracellular contents into the brain [1,6]. On the other hand, apoptosis occurs several hours or days after the stroke onset, where apoptosis occurs in the penumbra and is typically not detrimental to the neighboring cells [1,4]. Due to this, resident microglia are activated by the dead cells causing damage to the surrounding area, which also contributes to an increase of dead brain cells [1,5,6].

Microglia are the assigned immune cells aimed at protecting brain cells from any damage [1,2,7]. Microglia reside within the normal brain as a ubiquitously distributed quiescent cell population which respond to changes within the system's micro-environment, in order to interact swiftly with diseases [1,8,9]. When microglia are activated their functions change, with increased capacity for phagocytosis and production of cytokines and chemokines [1]. The small secreted proteins, cytokines, which are released by the cells produce a specific effect, especially on inter-cellular interactions and communications [1,5,6]. When cells go through the process of phagocytosis, they are able to identify and consume large particles like pathogens, apoptotic cells, and cellular debris [10].

In living systems, free radicals and oxidants have double functions as both toxic and beneficial complex since they can be either harmful or beneficial to the body [11–13]. The activation of microglia has both beneficial and deleterious effects. The positive effects manifest through the prevention of damage extension by phagocytosis and the production of trophic molecules and anti-inflammatory cytokines, which can be conducive to tissue repair as well as neuroprotection. The harmful effects, which are toxic to tissues, occur primarily through the production of free radicals [12,13]. This reaction is important in removing damaged cells from the brain tissue; however, it can also increase the harm to brain cells by producing free radicals that are toxic to healthy cells [2,14,15]. There has been an increasing amount of evidence suggesting that ischemic inflammation may be vital to the issue of pathogens, as ischemic stroke results not only in damage and neuronal loss but also in sustained neuroinflammation after stroke [3,16,17]. Microglial cells have been shown to take part in the neural analysis associated with adult brain function responsible for the harm occurring in the brain when a stroke occurs [3]. After an ischemic stroke, a dynamic microglial polarization occurs within the harm region [18]. The microglia composition M_1 unleashes proinflammatory cytokines and oxidative mediators that prejudice living neurons, whereas the microglia composition M_2 tends to unleash neurotrophic factors, in order to stop neuronal death and promote brain repair [17,18]. The anti-inflammatory phenotype composition M_2 is capable of manufacturing anti-inflammatory cytokines, which are conducive to inhibiting inflammation and tissue restoration [17–19]. This shows that microglia play both positive and negative roles in stroke [20,21].

In adult brains, neural stem cells (which are active throughout life) are capable of generating different cells [22,23]. Neurogenesis continues for the duration of life in the subventricular zone in all mammalian types, generating new neural stem cells during the recovery stage after ischemic stroke [24–27]. Neural stem cells appear to have many important roles, such as self-renewal and multipotency, replacement of the dead cells, and inhibition of inflammation [22,28–30]. However, endogenous neural stem cells may not be able to generate enough cells to repair the neurological damage caused by a major disease such as stroke [31]. Furthermore, the pathologic environment created after an ischemic stroke poses numerous hurdles for new neurons, which make the utilization of endogenous repair mechanisms more difficult; in particular, in directing the migration and differentiation of endogenous neural stem cells needed to repair the tissue [3,31].

Numerous mathematical models have been proposed for the dynamics of brain diseases, the functions of immune cells in terms of inflammation, and neural stem cell generation. Several models of inflammatory processes have been proposed. For example, the dynamics of inflammation from a stroke were modeled by Di Russo et al. [1] where they studied the dynamics of the densities of cells dead by necrosis and apoptosis, activated and inactivated resident microglia and the proportion of neutrophils and macrophages in the tissue. Mathematical models of inflammation proposed by the authors; Reynolds et al. [32] and Kumar et al. [33] ; for blood cells inflammation. Alharbi and Rambely [34] discussed a mathematical model of the ability of the immune system to inhibit and eliminate abnormal cells, as well as the role of dietary habits in boosting the immune system. Mathematical models for brain diseases have also been proposed by several authors [1,35]. For example, Hao and Friedman [35] described a model for Alzheimer's disease which utilized the microglia. Also, mathematical models have been proposed for compartmental adult neurogenesis. Nakata et al. [36] studied models of a hierarchical cell building system controlled by the mature cells. Ziebell et al. [37] introduced a mathematical model that represents different states of the adult hippocampus and the changing dynamics in stem cells during that time. Cacao and Cucinotta [38] developed a mathematical model of radiation-induced changes to neurogenesis, while a different study by Huang and Zhang [39] discussed the knowledge of strategies and mechanisms for neural stem cell-based therapies on brain hypoxic-ischemic injury. These studies were all focused on brain diseases and the roles of the immune system in several diseases in the blood or brain. Furthermore, some of the studies focused on the behavior of neural stem cell generation in the subventricular zone. As a starting point, we considered the model which has been recently proposed by Leah et al. [40]. We modified our models based on the model presented by Leah et al. [40], which was derived by studying the behaviors of several types of cells, microglia, brain cells, and the impact of microglial pro-inflammatory factors in living brain cells. In their model, they studied methods to represent interactions between proinflammatory and anti-inflammatory cytokines, types of microglia, and central nervous system (CNS) tissue damage, using clinical data [40].

In our model, we incorporated variables for several cytokines. Furthermore, we introduced the effect of the proinflammation by microglia on brain cells and focused on the brain during the time of stroke. However, we did not consider many types of cytokines. We studied the cytokines as parameters and their influence on the brain. Furthermore, we developed a second model to study the dynamics and stability of the interaction between the damage from microglia and the endogenous neural stem cells in the brain in the recovery stage, in order to study the ability of neural stem cells to heal the brain after stroke. The purposes of this work are to describe the biological interaction between the inflammatory cytokines from microglia and living brain cells in stroke; to study the damage on brain cells during the early stage of stroke over a 72-h duration; to study the interactions between microglia and neural stem cells and the influence of proinflammatory and anti-inflammatory cytokines on brain cells; to analyze the dynamic effect of microglia by stimulating the generation and the proliferation of neural stem cells during the recovery phase after the stroke; and to draw conclusions about the dynamics of neural stem cells to improve the brain after a stroke within a mathematical framework. Our final aim is to understand the positive and negative aspects of this biological process, which could be helpful for the development of therapeutic methods using endogenous neural stem cells in ischemic stroke, by studying the stability of the models.

In Section 2, we present a model—called stroke-microglia-damage (SMD)—of the effect of microglia on the brain during stroke onset, and another model—called stroke-microglia-neural stem cells-recovery (SMNR)—of the interaction between the microglia and neural stem cells and their impact on the brain during a stroke, including analysis of the equilibrium points of the models and their stability states. Numerical experiments of the modified models are detailed in Section 3. Finally, Section 4 presents the conclusions of the study.

2. Mathematical Models

We modified mathematical models of biological processes to investigate the impact of microglia and neural stem cells on the brain by increasing damage or assisting it in the recovery stage in two steps, illustrated by systems of ordinary differential equations. We developed the SMD model based on a model presented by Leah et al. [40], who proposed a mathematical model for neuroinflammation in traumatic brain injury using mathematical modeling with clinical data. The system of ordinary differential equations they derived described the dynamics of biological processes for multiple interactions post-injury. The first model, called the SMD model, that we shall develop describes the roles of microglia in the first 72 h after stroke onset without any generation of neural stem cells, while the second model, called the SMNR model, describes the state from the third day when neural stem cells start to generate. The resting microglia are the immune cells, M_0, which reside in the brain, and will be activated when a stroke occurs to a classic proinflammatory M_1 or an alternative anti-inflammatory M_2 state, where the resting microglia polarizes into two activation states, M_1 or M_2, in reaction to the dead cells from ischemia [18,19]. The cytokines signal for the activation of resting microglia into two phenotypes, proinflammation M_1, and anti-inflammation M_2, microglia denoted by R_1 and R_2, respectively. When microglia cells are activated in stroke, they become either M_1 or M_2

phenotypes. At the start of the stroke, the microglia are biased towards the M_1 phenotype rather than M_2 phenotype, since $R_1 > R_2$. We assume that in stroke onset with this bias towards M_1, the transition rate from M_1 into M_2 phenotype is very small, approximately equal to zero in the first 60 h after stroke onset [1], and the rate of anti-inflammation cytokines, R_2, is at a smaller rate than the rate of proinflammation cytokines, R_1, in the SMD model. We take $R_1 < R_2$ when the rate of proinflammation cytokines, R_1, is at a lower rate than the rate of anti-inflammation cytokines, R_2 in the recovery stage in the SMNR model. Neural stem cells in the subventricular zone from the adult mammalian brain give rise to neuroblasts, which migrate to the injury region and generate new cells [22,28,41]. Microglia have a positive and negative impact on generating neural stem cells [42]. M_1 can impair neural stem cells [18], while M_2 shows the ability to stimulate the generation and the proliferation of neural stem cells N_{sc} Ref. [19]. We focus our model on the mature neural stem cells only in the SMNR because no generation of neural stem cells occurs in this stage of stroke onset.

2.1. Modelling of the Effect of Microglia on the Brain in a Stroke Onset (SMD)

In this study, we modified the dynamic system of the SMD model based on the model presented in [40], which illustrates the behavior of the impact of activated microglia on brain cells in the 72 h following a stroke. We consider a non-linear system of ordinary differential equations, which describes the effects of microglia $M_0(t)$, $M_1(t)$, $M_2(t)$, and the damage from microglia, $D(t)$, on brain cells, $C(t)$, in the 72 h after stroke onset as follows:

$$\frac{dM_0}{dt} = \alpha - [R_1 + R_2 + \mu] M_0(t).$$

This equation demonstrates that the microglia are activated after an ischemic stroke and are polarized either towards a classic state, M_1, by proinflammatory cytokines, leading to an adaptive immune response and causing additional neuronal damage; or towards an alternative state, M_2, which is the anti-inflammatory phenotype induced by anti-inflammatory cytokines, which is thought to inhibit inflammation and enhance tissue repair [18,19,40]. Although we do not specify the types of cytokines in our models, we focus on the influences of the cytokines which gather in the region of damage; which are determined to either increase M_1 polarization or to change the microglia into the M_2 state. For mathematical modeling purposes, we assume that resting microglia M_0 are generated at a constant rate (α) and die at a constant rate (μ). The function of the activated microglia is to clean up dead cells produced by ischemia and cytokines caused by dead cells [43]. Microglia are activated by the cytokines caused by dead cells: the M_1 phenotype is induced by proinflammation cytokine signals R_1 and the M_2 phenotype is induced by anti-inflammation cytokine signals R_2 [5,17,43]. Furthermore, a shift from M_1 to M_2 may be induced by R_3 signaling [5], where R_3 is the result of a transition from proinflammatory to anti-inflammatory cytokines. This transition only appears in the SMNR model, occurring 60 h after the stroke onset [1]. The following differential equations illustrate the behavior of the activated microglia M_1 and M_2:

$$\frac{dM_1}{dt} = R_1 M_0(t) - [\delta D(t) + \gamma_1] M_1(t),$$
$$\frac{dM_2}{dt} = R_2 M_0(t) - \gamma_2 M_2(t),$$

where R_1 represents the rate of M_1 activation, R_2 is the rate of M_2 activation, δ is the rate of damage by M_1 activation, γ_1 is the death rate of M_1, and γ_2 is the death rate of M_2. The M_1 phenotype is essential for cell recovery, due to their protection mechanisms against the damage which recruits immune cells to the region of injury. The M_2 phenotype microglia also play a role in reducing damage, by clearing the brain of dead cells and assisting in neurogenesis, as well as inhibiting inflammation. On the other hand, the microglia can also play a negative role, through secretion of damaging proinflammatory cytokines which increase inflammation in the area of healthy cells [5,17]. We consider the damage D caused by the activated microglia to represent secondary damage in the case of stroke [2,14]. The following differential equation describes this damage:

$$\frac{dD}{dt} = D(t)[M_1(t)(\delta - r_1) - \beta_1 C(t) - r_2 M_2(t)],$$

where β_1 denotes the rate of effect of damage on brain cells, r_1 denotes the rate of damage clearance by M_1, and r_2 denotes the rate of damage clearance by M_2. In our model, we use the impact of the damage by microglia on the living brain cells, C. The following ordinary differential equation describes this effect on brain cells:

$$\frac{dC}{dt} = C(t)[\beta_1 D(t) - \beta_0],$$

where β_0 is the rate of cells dying from ischemic stroke.

Thus, the SMD model is expressed as follows:

$$\frac{dM_0}{dt} = \alpha - [R_1 + R_2 + \mu]M_0(t), \qquad (1)$$

$$\frac{dM_1}{dt} = R_1 M_0(t) - [\delta D(t) + \gamma_1]M_1(t), \qquad (2)$$

$$\frac{dM_2}{dt} = R_2 M_0(t) - \gamma_2 M_2(t), \qquad (3)$$

$$\frac{dD}{dt} = D(t)[M_1(t)(\delta - r_1) - \beta_1 C(t) - r_2 M_2(t)], \qquad (4)$$

$$\frac{dC}{dt} = C(t)[\beta_1 D(t) - \beta_0], \qquad (5)$$

with initial values $M_0(0) = 1$ [1], $M_1(0) = 0.1514$ [1], $M_2(0) = 0.02$ [35], $D(0) = 0.4$ [1], and $C(0) = 0.28$ [35]. Furthermore, we assumed that, in steady state, $M_1 > M_2$ for 72 h and that neural stem cells are not generated.

2.1.1. Equilibrium Points for the SMD Model

In this section, we calculate the equilibrium points of the system and determine the parameters for the existence of different types of biological states. We now determine the steady-state solutions as follows:

$$\frac{dM_0}{dt} = \frac{dM_1}{dt} = \frac{dM_2}{dt} = \frac{dD}{dt} = \frac{dC}{dt} = 0.$$

Equilibrium points are stable if they remain constant over time or continually balance change in one direction by that in another. We classify three steady states as follows:

Definition 1 (State of the activation of microglia). *We define the microglia that reside in a healthy brain as the absence of a high activation for these cells in the brain when any damage occurs in the region. The steady-state of the form $M_0; M_1; M_2 > 0$ and $D, C = 0$ indicates that the function of immune cells is normal and the microglia do not cause any damage in the brain.*

Definition 2 (State of the beginning of damage from activated microglia M_1). *We define the beginning of the damage by the increased rate of proinflammation cytokines where the existence of high activation of microglia will cause damage to the living brain cells. The steady-state of the form $M_0; M_1; M_2; D > 0, C = 0$ indicates the beginning activation of proinflammation from M_1 microglia.*

Definition 3 (State of the damage from activated microglia in brain cells). *We define the impact of the damage caused by microglia on the living brain cells. The steady state of the form $M_0; M_1; M_2; D; C > 0$ indicates damage of the living brain cells by active microglia.*

Populations of microglia cells (M_0, M_1, M_2), the damage on the tissue by proinflammation, D and the impact of this damage on the living brain cells in the CNS, C are positive or equal to zero for all equations. Thus, we obtain the equilibrium points by solving system (1)–(5) to determine the positive equilibrium points if and only if M_0, M_1, M_2, D and C exemplify the positive solutions.

Proposition 1 (Nonnegative Equilibrium for the SMD model). *We assume the equilibrium points for SMD system, $M_0; M_1; M_2; D; C > 0$ satisfy the following conditions:*

- $\delta > r_1$
- $R_1 \gamma_2(-r_1 + \delta) > r_2 R_2 \gamma_1$
- $\alpha R_1 \beta_1 \gamma_2(-r_1 + \delta) > r_2 R_2 (\beta_1 \gamma_1 + \beta_0 \delta)$.

Then and only then can there exist nonnegative real steady states.

Proof. The proof can be clearly found from Definition 1, 2 and 3. □

Hence, the equilibrium point is given as follows:
The equilibrium point for the activation of microglia, E_r

Corollary 1. *The equilibrium point $E_r = (M_0, M_1, M_2, 0, 0)$ exists in $\Re_+^5$ if and only if M_0, M_1, M_2 are the nonnegative roots and $D = 0, C = 0$. This point is given by*

$$E_r = (M_0, M_1, M_2, 0, 0) = \left(\frac{\alpha}{x}, \frac{\alpha R_1}{\gamma_1 x}, \frac{R_2 \alpha}{\gamma_2 x}, 0, 0 \right). \quad (6)$$

where,

$$x = R_1 + R_2 + \mu.$$

Thus, in the normal state in mammalian brains, we obtain a normal situation when there is no activation of microglia on the brain where $D, C = 0$ given that there is no damage on the brain cells, where the equilibrium point E_r indicated the activation of microglia steady state.
The equilibrium point for high activation of microglia, E_p

Corollary 2. *The equilibrium point E_p exists in $\Re_+^5$ if and only if, M_0, M_1, M_2 and D are the nonnegative roots and $C = 0$. Then, the equilibrium point of proinflammation microglia through 72 h after stroke is given as follows:*

$$E_p = \left(\frac{\alpha}{x}, \frac{r_2 R_2 \alpha}{\gamma_2(-r_1 + \delta)x}, \frac{R_2 \alpha}{\gamma_2 x}, \frac{-r_2 R_2 \gamma_1 + R_1 \gamma_2(-r_1 + \delta)}{r_2 R_2 \delta}, 0 \right). \quad (7)$$

The model has an equilibrium point for the activation of microglia. During this stage, M_1 has a significant impact. The equilibrium point E_p describes the activation by cytokines which obtains higher proinflammation from R_1 than the rate of the second kind of cytokines R_2 if and only if the rate of damage on the living brain cells by M_1 is more than r_1, which is the clearance damage by the immune cells microglia M_1.
The equilibrium points for the impact of proinflammation microglia M_1 on brain cells, E_d

Corollary 3. *The equilibrium point E_d exists in $\Re_+^5$ if and only if M_0, M_1, M_2, D and C are the nonnegative roots. Then, the equilibrium point E_d has an impact on the damage to the living brain cells through 72 h after stroke onset given as follows:*

$$E_d = \left(\frac{\alpha}{x}, \frac{R_1\beta_1\alpha}{\kappa x}, \frac{R_2\alpha}{\gamma_2 x}, \frac{\beta_0}{\beta_1}, \frac{(R_1\beta_1\gamma_2(-r_1+\delta) - r_2 R_2 \kappa)\alpha}{\beta_1 \gamma_2 \kappa x}\right), \quad (8)$$

where,

$$\kappa = (\beta_1 \gamma_1 + \beta_0 \delta).$$

2.1.2. Stability of Equilibrium Points

We now study the stability of equilibrium points for the effect of microglia on brain cells for 72 h without neural stem cells by using definitions 1–3. For the eigenvalues associated without neural stem cell equilibrium in stroke onset, the (5×5) Jacobian matrix of the system (1)–(5) is given by

$$J_1 = \begin{bmatrix} -x & 0 & 0 & 0 & 0 \\ R_1 & -\gamma_1 - D\delta & 0 & -M_1\delta & 0 \\ R_2 & 0 & -\gamma_2 & 0 & 0 \\ 0 & D(-r_1+\delta) & -Dr_2 & -M_2 r_2 - c\beta_1 + M_1(-r_1+\delta) & -D\beta_1 \\ 0 & 0 & 0 & c\beta_1 & D\beta_1 - \beta_0 \end{bmatrix}.$$

1. Stability analysis of equilibrium point, E_r:

   Theorem 1. *Suppose that the function $f : \Gamma \to \Re_+^5$ where Γ is a domain in $\Re_+^5$, and suppose that $E_r = (M_0, M_1, M_2, 0, 0) \in \Gamma$ is an equilibrium point at which at least one eigenvalue of the Jacobian matrix has a positive real part. Then, E_r is an unstable equilibrium point of f.*

   Proof. The Jacobian J_1 at equilibrium point E_r is calculated as follows:

   $$J[E_r] = \begin{bmatrix} -x & 0 & 0 & 0 & 0 \\ R_1 & -\gamma_1 & 0 & a_{24} & 0 \\ R_2 & 0 & -\gamma_2 & 0 & 0 \\ 0 & 0 & 0 & a_{44} & 0 \\ 0 & 0 & 0 & 0 & -\beta_0 \end{bmatrix},$$

   where,

   $$a_{24} = \frac{-R_1 \alpha \delta}{\gamma_1 x}, \quad a_{44} = \frac{-r_2 R_2 \alpha}{\gamma_2 x} + \frac{R_1 \alpha(-r_1 + \delta)}{\gamma_1 x}.$$

   The characteristic equation for the Jacobian $J[E_r]$ is given by

   $$(\beta_0 + \lambda)(\gamma_2 + \lambda)(\gamma_1 + \lambda)(x + \lambda)(\lambda + y) = 0. \quad (9)$$

   We assume that $x = R_1 + R_2 + \mu, y = \frac{r_2 R_2 \alpha}{\gamma_2 x} + \frac{R_1 \alpha(r_1 - \delta)}{\gamma_1 x}$.

   Then, the eigenvalues of Jacobian matrix $J[E_r]$ are given by:

   $$\lambda_1 = -x < 0, \quad \lambda_2 = -\beta_0 < 0,$$
   $$\lambda_3 = -\gamma_1 < 0, \quad \lambda_4 = -\gamma_2 < 0, \quad \lambda_5 = -y = \frac{\alpha(-r_2 R_2 \gamma_1 + R_1 \gamma_2(-r_1 + \delta))}{(\gamma_1 \gamma_2 x)} > 0.$$

The eigenvalues $\lambda_1, \ldots, \lambda_4$ are negative, but λ_5 is positive. Therefore, E_r is an unstable equilibrium point. □

2. Stability Analysis of Equilibrium point, E_p:

Theorem 2. *Suppose that the function $f : \Gamma \to \Re_+^5$ where Γ is a domain in $\Re_+^5$, and suppose that $E_p = (M_0, M_1, M_2, D, 0) \in \Gamma$ is an equilibrium point at which at least one eigenvalue of the Jacobian matrix has a positive real part and $D > 0$, $C = 0$. Then, E_p is an unstable equilibrium point of f.*

Proof. The Jacobian J_1, at E_p calculated as:

$$J[E_p] = \begin{bmatrix} -x & 0 & 0 & 0 & 0 \\ R_1 & b_{22} & 0 & b_{24} & 0 \\ R_2 & 0 & -\gamma_2 & 0 & 0 \\ 0 & b_{42} & b_{43} & b_{44} & b_{45} \\ 0 & 0 & 0 & 0 & b_{55} \end{bmatrix},$$

where,

$$b_{22} = -\gamma_1 - \frac{(-r_2 R_2 \gamma_1 + R_1 \gamma_2(-r_1 + \delta))}{r_2 R_2} < 0, \quad b_{24} = \frac{r_2 R_2 \alpha \delta}{\gamma_2 (r_1 - \delta) x} < 0,$$

$$b_{42} = \frac{(-r_1 + \delta)(-r_2 R_2 \gamma_1 + R_1 \gamma_2(-r_1 + \delta))}{r_2 R_2 \delta} > 0, \quad b_{43} = \frac{-(-r_2 R_2 \gamma_1 + R_1 \gamma_2(-r_1 + \delta))}{R_2 \delta},$$

$$b_{44} = \frac{-r_2 R_2 \alpha}{\gamma_2 x} - \frac{r_2 R_2 \alpha(-r_1 + \delta)}{\gamma_2 (r_1 - \delta) x} < 0, \quad b_{45} = -\frac{\beta_1(-r_2 R_2 \gamma_1 + R_1 \gamma_2(-r_1 + \delta))}{r_2 R_2 \delta},$$

$$b_{55} = \frac{R_1 \beta_1 \gamma_2(-r_1 + \delta) - r_2 R_2(\beta_1 \gamma_1 + \beta_0 \delta)}{r_2 R_2 \delta} > 0.$$

From the Jacobian $J[E_p]$, the characteristic equation is given by

$$(\gamma_2 + \lambda)(b_{55} - \lambda)(x + \lambda)(q_2 \lambda^2 + q_1 \lambda^1 + q_0) = 0,$$

where,

$$q_2 = r_2 R_2 \gamma_2 (r_1 - \delta) x, \quad q_1 = -R_1 \gamma_2^2 (r_1 - \delta)^2 x,$$
$$q_0 = -r_2 R_2 \alpha (r_2 R_2 \gamma_1 + R_1 \gamma_2 (r_1 - \delta))(r_1 - \delta),$$
$$x = R_1 + R_2 + \mu.$$

By Proposition 1 one of the eigenvalues is positive. So, $J(E_p)$ has one at least positive root, which indicates that the equilibrium point E_p is unstable [44]. □

3. Stability analysis of equilibrium point, E_d:

Theorem 3. *Suppose that the function $f : \Gamma \to \Re_+^5$ where Γ is a domain in $\Re_+^5$, and suppose that $E_d = (M_0, M_1, M_2, D, C) \in \Gamma$ is an equilibrium point where all the eigenvalues of the Jacobian matrix have negative real parts at the equilibrium point E_d and $D > 0$, $C > 0$. Then, E_d is a stable equilibrium point of f.*

Proof. The Jacobian J_1 at E_d is calculated as follows:

$$J[E_d] = \begin{bmatrix} -x & 0 & 0 & 0 & 0 \\ R_1 & c_{22} & 0 & c_{24} & 0 \\ R_2 & 0 & -\gamma_2 & 0 & 0 \\ 0 & c_{42} & c_{43} & c_{44} & -\beta_0 \\ 0 & 0 & 0 & c_{54} & 0 \end{bmatrix},$$

where,

$$c_{22} = -\gamma_1 - \frac{\beta_0 \delta}{\beta_1} < 0, \quad c_{24} = -\frac{R_1 \alpha \beta_1 \delta}{(\beta_1 \gamma_1 + \beta_0 \delta)x} < 0,$$

$$c_{42} = \frac{\beta_0(-r_1 + \delta)}{\beta_1} > 0, \quad c_{43} = -\frac{r_2 \beta_0}{\beta_1} < 0,$$

$$c_{44} = -\frac{r_2 R_2 \alpha}{\gamma_2 x} + \frac{R_1 \alpha \beta_1(-r_1 + \delta))}{(\beta_1 \gamma_1 + \beta_0 \delta)x}$$
$$+ \frac{\alpha(R_1 \beta_1 \gamma_2 (r_1 - \delta) + r_2 R_2 (\beta_1 \gamma_1 + \beta_0 \delta))}{\gamma_2 (\beta_1 \gamma_1 + \beta_0 \delta)x} > 0, \quad c_{45} = -\beta_0 < 0$$

$$c_{54} = -\frac{\alpha(R_1 \beta_1 \gamma_2 (r_1 - \delta) + r_2 R_2 (\beta_1 \gamma_1 + \beta_0 \delta))}{\gamma_2 (\beta_1 \gamma_1 + \beta_0 \delta)x} > 0.$$

The characteristic equation is given by

$$(\gamma_2 + \lambda)(x + \lambda)(N_3 \lambda^3 + N_2 \lambda^2 + N_1 \lambda + N_0) = 0. \tag{10}$$

Thus, we can find the first two eigenvalues directly:

$$\lambda_1 = -\gamma_2, \quad \lambda_2 = -x.$$

Here, we can apply the Routh–Hurwitz Criterion if and only if [45]:

- $N_2 > 0$,
- $N_0 > 0$,
- $N_2 N_1 > N_0 N_3$.

where

$$\begin{aligned} N_0 &= \alpha \beta_1 \beta_0 (\beta_1 \gamma_1 + \beta_0 \delta)(R_1 \beta_1 \gamma_2 (-r_1 + \delta) - r_2 R_2 (\beta_1 \gamma_1 + \beta_0 \delta)) > 0, \\ N_1 &= \alpha \beta_1^2 \beta_0 (R_1 \gamma_2 (-r_1 + \delta)(\beta_1 + \delta) - r_2 R_2 (\beta_1 \gamma_1 + \beta_0 \delta)) > 0, \\ N_2 &= \beta_1 \gamma_2 (\beta_1 \gamma_1 + \beta_0 \delta)^2 x > 0, \, N_3 = \beta_1^2 \gamma_2 (\beta_1 \gamma_1 + \beta_0 \delta)x > 0. \end{aligned}$$

Since from Proposition 1,

$$N_2 N_1 - N_0 N_3 = R_1 \alpha \beta_1^3 \beta_0 \gamma_2^2 \delta (-r_1 + \delta)(\beta_1 \gamma_1 + \beta_0 \delta)^2 x > 0.$$

Then, $N_2, N_0 > 0$ and $N_2 N_1 - N_0 N_3 > 0$,

Now, we apply the Routh–Hurwitz theorem for $N_3 \lambda^3 + N_2 \lambda^2 + N_1 \lambda + N_0 = 0$, giving

$$\begin{vmatrix} \lambda^3 & N_3 & N_1 \\ \lambda^2 & N_2 & N_0 \\ \lambda^1 & N^* & 0 \\ \lambda^0 & N_0 & 0 \end{vmatrix},$$

where,

$$N^* = \frac{N_2 N_1 - N_3 N_0}{N_2},$$

thus,

$$N^* = R_1 \alpha \beta_1^2 \beta_0 \gamma_2 \delta(-r_1 + \delta) > 0.$$

Since all the coefficients in the first column have positive signs; the Equation (10) has no roots with positive real parts and two of the eigenvalues are negative; thus, the equilibrium point E_d is stable. Activated microglia are capable of cleaning dead cells; however, they produce free radicals from brain cells, which increases the damage in brain cells during a stroke. This lead to further death of brain cells [17,32]. □

Remark 1. *The effect of microglia on brain cells in a stroke (which includes activated microglia at stroke onset) on the dynamic system of SMD model can be deduced, as follows:*

- *As a result of Theorem 1 and Definition 1, the damage, D, can invade the SMD system if $\lambda_5 > 0$.*
- *As a result of Theorem 2 and Definition 2, this means that the damage, $D > 0$, invades C.*
- *As a result of Theorem 3 and Definition 3, this means that the damage, $D > 0$, causes the death of C.*
- *The SMD model is stable when the brain cells are affected by the proinflammatory cytokines of microglia; however, when the rate of production of proinflammatory cytokines leads to an increase in damage, the possibility of death of the brain cells is introduced.*

2.2. Modeling the Interaction between Microglia and Neural Stem Cells and Impact on the Brain in Stroke (SMNR)

We formulate the following a non-linear system of ordinary differential equations to describe the interaction between microglial cells and neural stem cells and to investigate the damage to brain cells in the recovery stage after a stroke. This system uses the same equations as the SMD model, but also includes the transition from M_1 to M_2 and the behavior of neural stem cells N_{SC}. The system is as follows:

$$\frac{dM_0}{dt} = \alpha - [R_1 + R_2 + \mu] M_0(t) \tag{11}$$

$$\frac{dM_1}{dt} = R_1 M_0(t) - [R_3 + \delta D(t) + k_1 N_{SC}(t) + \gamma_1] M_1(t) \tag{12}$$

$$\frac{dM_2}{dt} = R_2 M_0(t) + R_3 M_1(t) - [k_2 N_{SC}(t) + \gamma_2] M_2(t) \tag{13}$$

$$\frac{dN_{SC}}{dt} = N_{SC}(t) [k_2 M_2(t) + k_1 M_1(t) - \beta_2 C(t) - \sigma] \tag{14}$$

$$\frac{dD}{dt} = D(t) [\delta M_1(t) - \beta_1 C(t) - r_1 M_1(t) - r_2 M_2(t)] \tag{15}$$

$$\frac{dC}{dt} = C(t) [\beta_1 D(t) + \beta_2 N_{SC}(t) - \beta_0], \tag{16}$$

with initial values $M_0[0] = 1$ [1], $M_1[0] = 0.02$ [35], $M_2[0] = 0.1514$ [1], $N_{SC}[0] = 0.85$ [38], $D[0] = 0.4$ Ref. [1], and $C[0] = 0.28$ [35].

The parameters are real and positive, where R_3 denotes the rate of transition from M_1 to M_2, k_1 is the rate of interaction between neural stem cells and M_1 microglia, k_2 is the rate of interaction between neural stem cells and M_2 microglia, β_2 is the rate of the effect of neural stem cells on brain cells, and σ is the death rate of neural stem cells. The other parameters are the same as in the SMD model. In the SMNR model, $M_1 < M_2$ occurs during the recovery stage, where neural stem cells also start to generate and help the brain. Therefore, microglial activation is important for directing the replacement of damaged or lost cells in the brain.

2.2.1. Equilibrium Points

In this section, we calculate the equilibrium points of the system (11)–(16) and determine the parameter conditions for the existence of the different types of biological states. We find the steady-state solutions as follows:

$$\frac{dM_0}{dt} = \frac{dM_1}{dt} = \frac{dM_2}{dt} = \frac{dN_{SC}}{dt} = \frac{dD}{dt} = \frac{dC}{dt} = 0.$$

We classify three states of steady states:

Definition 4 (The transformation state). *We define the transition cytokines from proinflammation microglia to anti-inflammation (M_1 to M_2). The steady state of the form M_0; M_1; $M_2 > 0$ and $N_{SC}, D, C = 0$ indicates the beginning of recovery stage.*

Definition 5 (The interaction between microglia and neural stem cells state). *We define the interaction between microglia and neural stem cells microglia in which the proinflammation phenotype (M_1) affects negatively on neural stem cells generation and the anti-inflammation phenotype (M_2) stimulates its generation. The steady state of the form M_0; M_1; M_2; $N_{SC} > 0$ and $C, D = 0$, indicates the interaction between brain cells in the recovery stage.*

Definition 6 (The impact of neural stem cells on brain cells state). *We define the impact of generating neural stem cells on brain cells, the steady state of the form M_0; M_1; M_2; N_{SC}; $C > 0$ and $D = 0$ indicates the generation of neural stem cells on brain cells.*

Populations of microglia cells (M_0, M_1, M_2), D, N_{SC} and C are positive or equal to zero. Thus, we obtain the equilibrium points by solving the system (11)–(16) to determine the positive equilibrium points if and only if, M_0, M_1, M_2, D, C exemplify the positive solution. Subpopulation of microglia cells M_0, M_1, M_2, N_{SC}, D and C are positive for all $t > 0$.

Proposition 2 (Nonnegative equilibriums for the SMNR model). *We assume the equilibrium points for the system SMNR is nonnegative for the following conditions given by:*

- $\delta > r_1$
- $\alpha\beta_2 > \beta_0\sigma$
- $R_2\alpha\beta_2 > \beta_0\sigma(R_2 + \mu)$
- $k_1\alpha\beta_2 > k_1\beta_0\sigma + \beta_2\gamma_1\sigma$
- $(k_1\beta_0(R_2 + \mu)\sigma + \beta_2(R_3 + \gamma_1)x\sigma < k_1R_1(\alpha\beta_2 - \beta_0\sigma)$
- $k_2\gamma_1 < k_1\gamma_2$
- $k_1k_2\alpha + k_2R_3\sigma + k_2\gamma_1\sigma > k_1\gamma_2\sigma$
- $k_1k_2(R_1 + R_2)\alpha + k_2(R_3 + \gamma_1)x\sigma < k_1\gamma_2x\sigma + \nu$
- $k_1k_2(R_1 + R_2)\alpha + k_2R_3x\sigma + k_1\gamma_2x\sigma > k_2\gamma_1x\sigma + \nu$
- $k_1k_2(R_1 + R_2)\alpha + \nu > k_1\gamma_2x\sigma + k_2(R_3 + \gamma_1)x\sigma.$

Then and only then there exist non-negative real steady states.

Proof. The proof is clearly found from the steady states imposing $(M_0; M_1; M_2; N_{SC}; D; C) > 0$. Thus the models (11)–(16) obtain three points of equilibrium in the recovery stage by using MATHEMATICA. □

The first equilibrium point: the transformation state of microglia from proinflammation to anti-inflammation, E_t

Corollary 4. *The equilibrium point E_t has transformation microglia from proinflammation to anti-inflammation point recovery state from stroke given as follows:*

The equilibrium point $E_t = (M_0, M_1, M_2, N_{SC}, D, C)$ exists in $\mathfrak{R}_+^6$, in the beginning of activation of microglia that inhibits and transfers part of M_1 phenomena to the second M_2 phenomena. Then, $C = 0$, $D = 0$, $N_{SC} = 0$. This point E_t is given by

$$E_t = \left(\frac{\alpha}{x}, \frac{(R_1\alpha)}{(R_3+\gamma_1)x}, \frac{\alpha(R_1R_3+R_2(R_3+\gamma_1))}{(R_3+\gamma_1)\gamma_2 x}, 0, 0, 0\right), \quad (17)$$

where

$$x = R_1 + R_2 + \mu.$$

The second equilibrium point: the interaction between microglia and neural stem cells on brain cells, E_I

Corollary 5. *The equilibrium point $E_I = (M_0, M_1, M_2, N_{SC}, D, C)$ exists in $\mathfrak{R}_+^6$. At the beginning of the neural stem cells generation, the damage decreases in this state. Then, $C = 0$, $D = 0$. This point is given by*

$$E_I = \left(\frac{\alpha}{x}, \frac{1}{2k_1(-k_2\gamma_1+k_1\gamma_2)x}[(R_1+R_2)(-k_1k_2\alpha-k_2R_3\sigma-k_2\gamma_1\sigma+k_1\gamma_2\sigma)\right.$$

$$-\mu\sigma[k_2(R_3+\gamma_1)-k_1\gamma_2]+v], \frac{1}{2k_2(-k_2\gamma_1+k_1\gamma_2)x}[(R_1+R_2)(k_1k_2\alpha$$

$$+k_2R_3\sigma-k_2\gamma_1\sigma+k_1\gamma_2\sigma)+\mu\sigma(k_2(R_3-\gamma_1)+k_1\gamma_2)]-v$$

$$,\frac{1}{2k_1k_2x\sigma}[(R_1+R_2)(k_1k_2\alpha-k_2R_3\sigma-k_2\gamma_1\sigma$$

$$\left.-k_1\gamma_2\sigma)-k_2\mu\sigma(R_3+\gamma_1)-k_1\gamma_2\mu\sigma+v), 0, 0\right) \quad (18)$$

where,

$$v = \sqrt{(4k_1k_2R_1\alpha(-k_2\gamma_1+k_1\gamma_2)x\sigma+(k_1k_2(R_1+R_2)\alpha+k_2(R_3+\gamma_1)x\sigma-k_1\gamma_2 x\sigma)^2)}.$$

The third equilibrium point: the effect of generating neural stem cells on brain cells, E_R

Corollary 6. *The equilibrium point $E_R = (M_0, M_1, M_2, N_{SC}, D, C)$ exists in of $\mathfrak{R}_+^6$. When the neural stem cells are generated, the damage fades away in this state. Then, $D = 0$. This point is given by:*

$$E_R = \left(\frac{\alpha}{x}, \frac{R_1\alpha\beta_2}{\epsilon_1 x}, \frac{\alpha\beta_2(R_1R_3\beta_2+R_2\epsilon_1)}{\epsilon_1\epsilon_2 x}, \frac{\beta_0}{\beta_2}, 0\right.$$

$$\left., \left(\frac{\beta_2\gamma_2(-k_1\beta_0(R_2+\mu)\sigma-\beta_2(R_3+\gamma_1)x\sigma+k_1R_1\epsilon_3)+k_2\epsilon_7}{\beta_2\epsilon_1\epsilon_2 x}\right)\right), \quad (19)$$

where,

$$\begin{aligned}
\epsilon_1 &= (R_3\beta_2 + k_1\beta_0 + \beta_2\gamma_1), \quad \epsilon_2 = k_2\beta_0 + \beta_2\gamma_2, \\
\epsilon_3 &= (\alpha\beta_2 - \beta_0\sigma), \quad \epsilon_4 = (-k_1\alpha\beta_2 + k_1\beta_0\sigma + \beta_2\gamma_1\sigma), \\
\epsilon_5 &= R_3\beta_2 + k_1\beta_0 + \beta_2\gamma_1, \quad \epsilon_6 = (R_2\alpha\beta_2 - R_2\beta_0\sigma - \beta_0\mu\sigma), \\
\epsilon_7 &= R_1R_3\beta_2\epsilon_3 - R_1\beta_0\epsilon_4 + \epsilon_5\epsilon_6.
\end{aligned}$$

2.2.2. Stability of Equilibrium Points for SMNR Model

We study the stability of the equilibrium points for the second model SMNR by using Definitions 4–6. The first step requires linearization of the system equations that describes the interaction between microglia M_1, M_2, neural stem cells N_{SC}, the damage D in this stage and the impact it has on the brain cells C. The (6×6) Jacobian matrix of the system (11)–(16) is given by:

$$J_2 = \begin{bmatrix}
j_{11} & 0 & 0 & 0 & 0 & 0 \\
R_1 & j_{22} & 0 & -k_1M_1 & -M_1\delta & 0 \\
R_2 & R_3 & j_{33} & -k_2M_2 & 0 & 0 \\
0 & k_1N_{SC} & k_2N_{SC} & j_{54} & 0 & -N_{SC}\beta_2 \\
0 & D(-r_1+\delta) & -Dr_2 & 0 & j_{55} & -D\beta_1 \\
0 & 0 & 0 & c\beta_2 & c\beta_1 & j_{66}
\end{bmatrix},$$

where,

$$\begin{aligned}
j_{11} &= -x, \, x = R_1 + R_2 + \mu, \quad j_{22} = -R_3 - k_1N_{SC} - \gamma_1 - D\delta, \\
j_{33} &= -k_2N_{SC} - \gamma_2, \quad j_{54} = k_1M_1 + k_2M_2 - C\beta_2 - \sigma, \\
j_{55} &= -M_2r_2 - C\beta_1 + M_1(-r_1+\delta), \quad j_{66} = D\beta_1 + N_{SC}\beta_2 - \beta_0.
\end{aligned}$$

1. Stability analysis of equilibrium point, E_t:

Theorem 4. *Suppose that the function $f : \Gamma \to \Re_+^6$ where Γ is a domain in $\Re_+^6$, and suppose that $E_t = (M_0, M_1, M_2, 0, 0, 0) \in \Gamma$ is an equilibrium point at which at least one eigenvalue of the Jacobian matrix has a positive real part. Then, E_t is an unstable equilibrium point of f.*

Proof. The Jacobian J_2 corresponding to the equilibrium point E_t is given by

$$J[E_t] = \begin{bmatrix}
-x & 0 & 0 & 0 & 0 & 0 \\
R_1 & -R_3-\gamma_1 & 0 & d_{24} & d_{25} & 0 \\
R_2 & R_3 & -\gamma_2 & d_{34} & 0 & 0 \\
0 & 0 & 0 & d_{44} & 0 & 0 \\
0 & 0 & 0 & 0 & d_{55} & 0 \\
0 & 0 & 0 & 0 & 0 & -\beta_0
\end{bmatrix},$$

where,

$$\begin{aligned}
d_{24} &= -\frac{k_1R_1\alpha}{(R_3+\gamma_1)x}, \quad d_{25} = -\frac{R_1\alpha\delta}{(R3+\gamma_1)x}, \\
d_{34} &= -\frac{k_2\alpha(R_1R_3 + R_2(R_3+\gamma_1))}{(R_3+\gamma_1)\gamma_2 x}, \quad d_{44} = \frac{k_1R_1\alpha}{(R_3+\gamma_1)x} + \frac{k_2\alpha(R_1R_3 + R_2(R_3+\gamma_1))}{(R_3+\gamma_1)\gamma_2 x} - \sigma, \\
d_{55} &= -\frac{r_2\alpha(R_1R_3 + R_2(R_3+\gamma_1))}{(R_3+\gamma_1)\gamma_2 x} + \frac{R_1\alpha(-r_1+\delta)}{(R_3+\gamma_1)x}.
\end{aligned}$$

From the Jacobian matrix $J[E_t]$, the characteristic equation is given by

$$(\beta_0 + \lambda)(R_3 + \gamma_1 + \lambda)(\gamma_2 + \lambda)(x + \lambda)(-\lambda + t_{55})(-\lambda + t_{44}) = 0. \quad (20)$$

Then, the eigenvalues corresponding to E_t are given by

$$\lambda_1 = -\beta_0 < 0, \quad \lambda_2 = -R_3 - \gamma_1 < 0, \quad \lambda_3 = -x < 0, \quad \lambda_4 = -\gamma_2 < 0,$$

$$\lambda_5 = \frac{k_1 R_1 \alpha}{(R_3 + \gamma_1)x} + \frac{k_2 \alpha (R_1 R_3 + R_2(R_3 + \gamma_1))}{(R_3 + \gamma_1 \gamma_2)x} - \sigma > 0,$$

$$\lambda_6 = -\frac{r_2 \alpha (R_1 R_3 + R_2(R_3 + \gamma_1))}{(R_3 - \gamma_1 \gamma_2)x} + \frac{(R_1 \alpha (r_1 - \delta))}{(R_3 + \gamma_1)x} < 0.$$

One of the eigenvalues, $\lambda_5 > 0$, then E_t is an unstable point. □

2. Stability analysis of equilibrium point, E_I:

Theorem 5. *Suppose that the function* $f : \Gamma \to \Re_+^6$ *where* Γ *is a domain in* $\Re_+^6$, *and suppose that* $E_I = (M_0, M_1, M_2, N_{SC}, 0, 0) \in \Gamma$ *is an equilibrium point at which at least one eigenvalue of the Jacobian matrix has a positive real part and* $N_{SC} > 0$. *Then,* E_I *is an unstable equilibrium point of* f.

Proof. The Jacobian matrix J_2 corresponding to the equilibrium point E_I is given by:

$$J[E_I] = \begin{bmatrix} -x & 0 & 0 & 0 & 0 & 0 \\ R_1 & e_{22} & 0 & e_{24} & e_{25} & 0 \\ R_2 & R_3 & e_{33} & e_{34} & 0 & 0 \\ 0 & e_{42} & e_{43} & 0 & 0 & e_{46} \\ 0 & 0 & 0 & 0 & e_{55} & 0 \\ 0 & 0 & 0 & 0 & 0 & e_{66} \end{bmatrix},$$

where,

$$e_{22} = -\frac{\Lambda_1 + v}{2k_2 x \sigma} < 0, \quad e_{42} = \frac{\Lambda_2}{2k_2 x \sigma} > 0, \quad e_{33} = -\frac{\Lambda_2}{2k_1 x \sigma} < 0, \quad e_{43} = \frac{\Lambda_2}{2k_1 x \sigma} > 0,$$

$$e_{24} = \frac{-\Lambda_1 + v}{2qx} < 0, \quad e_{34} = \frac{\Lambda_3}{2qx} < 0, \quad e_{25} = -\frac{\delta \Lambda_4}{2k_1 qx} < 0,$$

$$e_{46} = \frac{\Lambda_7}{2k_1 k_2 x \sigma} < 0, \quad e_{55} = -\frac{\Lambda_5}{2k_1 k_2 qx} > 0, \quad e_{66} = \frac{\Lambda_6}{2k_1 k_2 x \sigma} > 0,$$

$$v = \sqrt{4k_1 k_2 R_1 \alpha (-k_2 \gamma_1 + k_1 \gamma_2) x \sigma + (k_1 k_2 (R_1 + R_2) \alpha + k_2 (R_3 + \gamma_1) x \sigma - k_1 \gamma_2 x \sigma)^2} > 0,$$

$$\Psi = x(k_2(R_3 + \gamma_1) + k_1 \gamma_2)\sigma > 0, \Lambda_1 = k_1 k_2 (R_1 + R_2) \alpha + k_2 x (R_3 + \gamma_1) \sigma - k_1 x \gamma_2 \sigma > 0,$$

$$\Lambda_2 = v - \Psi + k_1 k_2 (R_1 + R_2) \alpha > 0,$$

$$\Lambda_3 = -v + k_1 k_2 (R_1 + R_2) \alpha + k_2 x (R_3 - \gamma_1) \sigma + k_1 x \gamma_2 \sigma > 0,$$

$$\Lambda_4 = -v + k_1 k_2 (R_1 + R_2) \alpha + (k_2 (R_3 + \gamma_1) - k_1 \gamma_2) x \sigma < 0, \Lambda_5 = k_2^2 (r_1 - \delta)(k_1 (R_1 + R_2) \alpha$$
$$+ x(R_3 + \gamma_1) \sigma) + k_1 r_2 (v - k_1 x \gamma_2 \sigma) + k_2 (-k_1 r_2 (k_1 (R_1 + R_2) \alpha + x (R_3 - \gamma_1) \sigma)$$
$$- (r_1 - \delta)(v + k_1 x \gamma_2 \sigma)) > 0,$$

$$\Lambda_6 = v \beta_2 + k_1 k_2 (R_1 + R_2) \alpha \beta_2 - x(k_2 \beta_2 (R_3 + \gamma_1) + k_1 (2k_2 \beta_0 + \beta_2 \gamma_2)) \sigma > 0,$$

$$\Lambda_7 = \beta_2 (-v - k_1 k_2 (R_1 + R_2) \alpha + k_2 x (R_3 + \gamma_1) \sigma + k_1 x \gamma_2 \sigma) < 0,$$

$$q = k_2 \gamma_1 - k_1 \gamma_2 < 0, x = R_1 + R_2 + \mu > 0.$$

From the Jacobian $J[E_I]$, the characteristic equation is given by

$$(e_{66} - \lambda)(\lambda + x)(e_{55} - \lambda)[\lambda^3 + T_2\lambda^2 + T_1\lambda + T_0] = 0, \tag{21}$$

where,

$$\begin{aligned}
T_3 &= -8k_1^3 k_2^2 q^2 x^4 \sigma^3, \quad T_2 = -4k_1^2 k_2 q^2 (k_1(v + \Lambda_1) + k_2\Lambda_2) x^3 \sigma^2, \\
T_1 &= 2k_1^2 k_2 q x^2 \Lambda_2 \sigma(-q(v + \Lambda_1) + \Lambda_2(k_1(v - \Lambda_1) + k_2\Lambda_3)\sigma), \\
T_0 &= k_1^2 k_2 q \Lambda_2 x \sigma(\Lambda_1(-\Lambda_2 + \Lambda_3 - 2k_2 R_3 x \sigma) + v(\Lambda_2 + \Lambda_3 + 2k_2 R_3 x \sigma)).
\end{aligned}$$

In Equation (21), the eigenvalue $\lambda = e_{66}$ is distinctly positive, by Proposition 2. Thus, $J(E_I)$ has at least one positive root. Thus, the equilibrium point E_I is unstable [44]. □

3. Stability analysis of equilibrium point, E_R:

Theorem 6. *Suppose that the function $f : \Gamma \to \Re_+^6$ where Γ is a domain in $\Re_+^6$, and suppose that $E_R = (M_0, M_1, M_2, N_{SC}, 0, C) \in \Gamma$ is an equilibrium point at which at least one eigenvalue of the Jacobian matrix has a positive real part and $N_{SC} > 0$, $C > 0$. Then, E_R is an unstable equilibrium point of f.*

Proof. We now study the stability of the equilibrium point E_R, calculated as:
The Jacobian matrix J_2 estimated at E_R is

$$J[E_R] = \begin{bmatrix} -x & 0 & 0 & 0 & 0 & 0 \\ R_1 & f_{22} & 0 & f_{24} & f_{25} & 0 \\ R_2 & R_3 & f_{33} & f_{34} & 0 & 0 \\ 0 & \frac{k_1\beta_0}{\beta_2} & \frac{k_2\beta_0}{\beta_2} & f_{44} & 0 & -\beta_0 \\ 0 & f_{52} & f_{53} & 0 & f_{55} & f_{56} \\ 0 & 0 & 0 & 0 & 0 & f_{66} \end{bmatrix},$$

where

$$\begin{aligned}
f_{22} &= -R_3 - \frac{k_1\beta_0}{\beta_2} - \gamma_1 - \frac{\delta\rho_1}{\beta_2 v_1 v_2 x} < 0, \quad f_{24} = -\frac{k_1 R_1 \alpha \beta_2}{v_1 x} < 0, \quad f_{25} = -\frac{(R_1\alpha\beta_2\delta)}{v_1 x} < 0, \\
f_{55} &= -\frac{(r_2\alpha\beta_2)\rho_2}{v_1 v_2 x} + \frac{R_1\alpha\beta_2(-r_1 + \delta)}{v_1 x} > 0, \quad f_{33} = -\frac{k_2\beta_0}{\beta_2} - \gamma_2, \quad f_{34} = -\frac{k_2\alpha\beta_2\rho_2}{v_1 v_2 x} > 0, \\
f_{43} &= \frac{k_2\beta_0}{\beta_2}, \quad f_{44} = \frac{k_1 R_1\alpha\beta_2}{v_1 x} + \frac{k_2\alpha\beta_2\rho_2}{v_1 v_2 x} - \sigma > 0, \quad f_{42} = \frac{k_1\beta_0}{\beta_2}, \quad f_{53} = -\frac{r_2\rho_1}{\beta_2 v_1 v_2 x} < 0, \\
f_{56} &= -\frac{\beta_1\rho_1}{\beta_2 v_1 v_2 x} < 0, \quad f_{52} = \frac{(-r_1 + \delta)\rho_1}{\beta_2 v_1 v_2 x} < 0, \quad f_{66} = \frac{\beta_1\rho_1}{\beta_2 v_1 v_2 x} > 0, \\
\rho_1 &= (\beta_2\gamma_2(-k_1\beta_0(R_2 + \mu)\sigma - \beta_2(R_3 + \gamma_1)x\sigma + k_1 R_1(\alpha\beta_2 - \beta_0\sigma)), \\
&\quad + k_2(R_1 R_3\beta_2(\alpha\beta_2 - \beta_0\sigma)) - R_1\beta_0(-k_1\alpha\beta_2 + k_1\beta_0\sigma + \beta_2\gamma_1\sigma) \\
&\quad + (R_3\beta_2 + k_1\beta_0 + \beta_2\gamma_1)(R_2\alpha\beta_2 - R_2\beta_0\sigma - \beta_0\mu\sigma))) \\
\rho_2 &= (R_1 R_3\beta_2 + R_2(R_3\beta_2 + k_1\beta_0 + \beta_2\gamma_1)) > 0, \\
v_1 &= (R_3\beta_2 + k_1\beta_0 + \beta_2\gamma_1) > 0, \quad v_2 = (k_2\beta_0 + \beta_2\gamma_2) > 0.
\end{aligned}$$

From the Jacobian $J[E_R]$, the characteristic equation is given by

$$(f_{66} - \lambda)(\lambda + x)(f_{25}f_{34}(-f_{53}k_1 + f_{52}k_2)\beta_0 + f_{25}\beta_2(f_{44} - \lambda)(-f_{33}f_{52} + f_{53}R_3 + f_{52}\lambda) + (f_{55} - \lambda)$$
$$\times((-f_{34}k_2\beta_0 + \beta_2(f_{33} - \lambda)(f_{44} - \lambda))(f_{22} - \lambda) + f_{24}\beta_0(-f_{33}k_1 + k_2 R_3 + k_1\lambda))) = 0. \tag{22}$$

In Equation (22), the eigenvalue $\lambda = f_{66}$ is distinctly positive, by Proposition 2. Thus, $J(E_R)$ has at least one positive root and, so, the equilibrium point E_R is unstable [44]. □

Remark 2. *The effects of microglia and neural stem cells on brain cells in the recovery stage after a stroke, according to analysis of the dynamic system of SMNR, can be deduced as follows:*

- As a result of Theorem 4 and Definition 4, the neural stem cells, Nsc, can invade the SMNR system if $\lambda_5 > 0$.
- As a result of Theorem 5 and Definition 5, this means that the neural stem cells, $Nsc > 0$, can invade C and D.
- As a result of Theorem 6 and Definition 6, this means that the neural stem cells, $Nsc > 0$, can eliminate the damage D.
- The SMNR model is unstable, given that the model uses mature neural stem cells, where the neural stem cells can help the brain to inhibit inflammation during a stroke; however, it is not always enough to substitute all dead cells with new neural stem cells [3,27].

3. Numerical Experiments

The aim of this section is to study the parameters of the systems (1)–(5) and (11)–(16), in order to determine those that affect the behavior of the modified models, by using numerical simulations. One of the main problems in modeling and simulating the interactions between microglia and endogenous neural stem cells is that few parameter values are known. Therefore, we listed the parameters which could be determined from the stroke model of [1]. The other parameters were determined from experimental data of other brain injuries which involve similar biological processes. Furthermore, we obtained some parameter values by simulations using the Mathematica software (11.2, Wolfram, Champaign, IL, USA) with the command NDSolve, in order to study the influence of the interactions during the stroke and the possible therapeutic value of neural stem cells. Tables 1 and 2 show the reference sets of parameter values, along with the corresponding simulation results.

Table 1. Parameters values for the stroke-microglia-damage (SMD) model.

The SMD Model			
Parameters	Values	Meaning	Sources
α	0.38	Source of resting microglia	[1]
R_1	0.12	Rate the activation of resting microglia into M_1	[1]
R_2	0.017	Rate the activation of resting microglia into M_2	[35]
δ	0.2854	Rate of the damage produced by M_1	[1]
β_1	0.1	Rate of effect the damage on $C(t)$	[1]
β_0	0.05	Rate of dying C through stroke	[1]
μ	0.003	Death rate of M_0	simulation
γ_1	0.05	Death rate of M_1	simulation
γ_2	0.06	Death rate of M_2	simulation
r_1	0.05	Damage clearance by M_1	[1]
r_2	0.0125	Damage clearance by M_2	[1]

The numerical simulations in Tables 1 and 2 appeared in the SMD and the SMNR models. The microglial cells at stroke onset had high proinflammatory activation (i.e., high level of M_1), which caused an increase of damage in the region, affecting brain cells as well as neural stem cells during the 72 h period. On the other hand, when the rate of the anti-inflammatory phenotype microglia (M_2) became higher than that of the M_1, the brain was assisted in recovery and the neural stem cells were stimulated to generate new cells to compensate for lost brain cells. These dynamics started approximately on the third day after stroke onset. The parameters of the SMD model, in the simulation, were fixed as $\mu = 0.003$, $\gamma_1 = 0.05$, and $\gamma_2 = 0.06$. The parameters of the SMNR model, in the simulation, were fixed as $R_3 = 0.11$, $K_1 = 0.75$, $K_2 = 0.91$, $\beta_2 = 0.2$, and $\mu = 0.053$. In the

SMD model, $M_1 > M_2$ and we set $R_1 = 0.12$ [1] and $R_2 = 0.017$ [35]. In the SMNR model, $M_2 > M_1$, the source of resting microglia was $\alpha = 0.38$, and the rate of the anti-inflammatory cytokines (R_2) was greater than that of the proinflammatory cytokines (R_1) in the recovery stage. Accordingly, we set $R_1 = 0.12$ [1] and estimated $R_2 = 0.26$. The dynamics of the systems were given by solving the systems numerically and then plotting the time series of the solutions of the system (1)–(5) and (11)–(16) for the parameters, which were solved using the fourth-order Runge–Kutta (RK4) method to obtain more stable and convergent solutions. The simulations of SMD and SMNR were carried out for the time spans of three days (72 h) and thirty days, respectively with a step size of 10^{-4} in the RK4 method. The SMD model described the inflammatory process where activation of the microglia cells peaks around three days (72 h) from stroke onset [1,2]. Neural stem cells increase the proliferation and the migration from the subventricular zone to the brain in thirty days and peaking around seven days after stroke [3,27]. Therefore, the SMNR model studied thirty days after stroke to investigate the interactions between microglia and neurogenesis after stroke. The accuracy of solutions by using the RK4 method of the SMD model and the best numerical solution of the model was obtained at 38 steps. While the accuracy of solutions by using the RK4 method of the SMNR model and the best numerical solution of the model was obtained at 35 steps. The reliability and accuracy of the proposed numerical method can be seen from the residual error which is shown in Figures 1 and 2. Figure 1 shows the residual error for the SMD model at the number of steps and the residual error at the time. While Figure 2 shows the residual error for the SMNR model at the number of steps and the residual errors at the time. By comparing the results in our study of microglial activation and the damage during the first three days after stroke onset and the recovery stage within the first thirty days after stroke onset with the effects of neural stem cells and those of the study in [40], we found that the studies agreed, in that the extended proinflammatory activation of microglia may inhibit neurogenesis and contribute to additional neuronal loss. Hence, the behavioural responses of microglia can lead to an increase in damage in the brain or its capability to recover. In this study, we focused on the importance and contribution of neural stem cells in transitioning to the recovery stage, as well as the interaction between microglia and neural stem cells. The generation of neural stem cells was considered to lead to a decrease in damage to the brain, compared to their absence. The behaviors of the solutions of the SMD and SMNR models are shown in Figure 3. It can be seen, in the SMD model, that the population of microglial cells M_1 had a shallow increase, asides from shifting to a steep curve at approximately 15–30 h, and peaked at around 72 h. On the other hand, we can observe that M_2 had a low level after stroke onset and, after three days, reached a stable level. The population of M_1 began to shift into the steep curve (the second stage) with an increase of M_2. Furthermore, the damage increased sharply in the initial stage and decreased in the second stage. On the other hand, in the SMNR model, the response of the neural stem cells began after the initial stage, helping the brain cells to repair damaged cells and lost cells, depending on whether the number of living cells was greater than that of the dead cells. As depicted in Figure 3, the curve of neural stem cells increased and then decreased during the 15 days following stroke onset; this can be explained by the ability of neural stem cells to inhibit inflammation, as has been studied in [3], where the survival of neural stem cells seemed to be prevented, with up to 80% of the new neurons dying within two weeks after their generation in vivo. Furthermore, the curve of neural stem cells in Figure 3 reached its peak during the first week, in agreement with studies in rats and mice, where the increase in proliferation of neural stem cells peaked at around 7 days after ischemic injury [27].

Table 2. Parameters values for the stroke-microglia-neural stem cells-recovery (SMNR) model.

The SMNR Model			
Parameters	Values	Biological Meaning	Sources
α	0.38	Source of resting microglia	[1]
R_1	0.12	Rate the activation of resting microglia into M_1	[1]
R_2	0.26	Rate the activation of resting microglia into M_2	estimated
R_3	0.11	The transition rate of $M_1 \to M_2$	simulation
K_1	0.91	Rate of interaction between M_1 and N_{SC}	simulation
K_2	0.75	Rate of interaction between M_2 and N_{SC}	simulation
δ	0.2854	Rate of the damage produced by M_1	[1]
β_1	0.1	Rate of effect the damage on $C(t)$	[1]
β_2	0.2	Rate of effect N_{SC} on $C(t)$	simulation
β_0	0.05	Rate of dying C through stroke	[1]
μ	0.053	Death rate of M_0	simulation
γ_1	0.015	Death rate of M_1	[35]
γ_2	0.015	Death rate of M_2	[35]
σ	0.015	Death rate of N_{SC}	[38]
r_1	0.05	Damage clearance by M_1	[1]
r_2	0.0125	Damage clearance by M_2	[1]

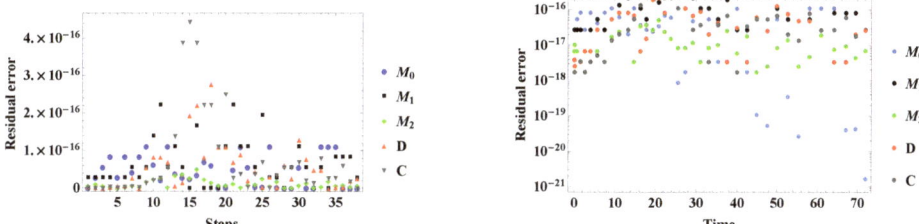

Figure 1. Residual error for the step and time of the numerical method in the SMD model.

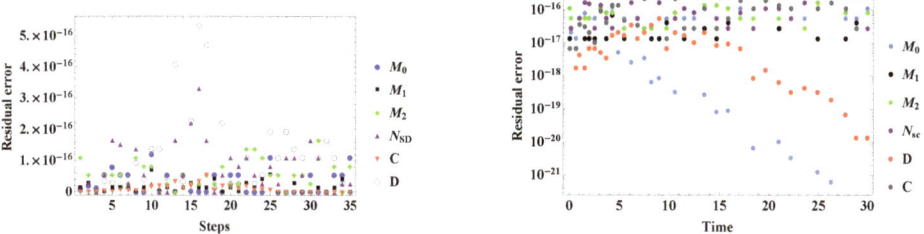

Figure 2. Residual error for the step and time of the numerical method in the SMNR model.

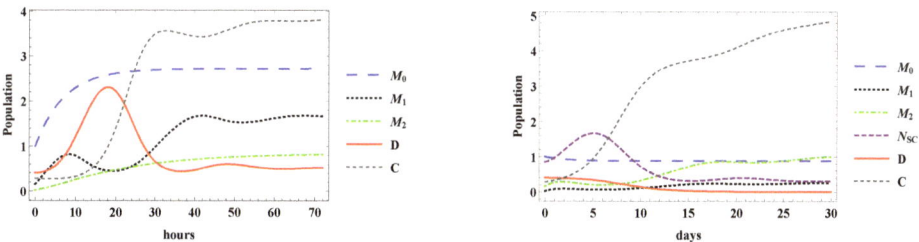

Figure 3. The behaviour of the SMD model within 72 h from stroke onset and the SMNR model within 30 days after stroke onset.

4. Conclusions

We modified two models—SMD ((1)–(5)) and SMNR ((11)–(16))—based on the model of Leah et al. [40], in order to study the impact of the microglial activation stage and the interaction between microglia and neural stem cells during a stroke. The dynamic models of the effects of microglia and the interaction between neural stem cells and microglia in stroke over time were studied both analytically and numerically. From the results of the analysis and simulation of both models, two states of microglial cells (M_1 to M_2) emerged from the resting state. In the first disease stage, the activation of microglial cells went from resting-state microglia to either of the states M_1 or M_2. The results of this interaction led to increased damage in the brain after stroke onset when M_2 was very low, there was no shift from M_1 to M_2, and no generation of neural stem cells occurred in the first 72 h after the stroke. Subsequently, in the recovery stage, the rate of M_1 decreased and the rate of M_2 increased and the neural stem cells began to generate. Subsequent to that, the rate of damage decreased. The instability of the SMNR model can be explained: in the nervous system, the number of endogenous neural stem cells is very low under normal physiological conditions, showing a very limited capacity for cell replacement under normal physiological conditions [20,21]. Understanding the biology in vivo of neural stem cells could lead to new therapeutic strategies for brain repair by endogenous neural stem cells [20,21], encouraging inflammatory inhibition. In conclusion, our modified models can lead to lead to an understanding of the effectiveness risks of the inflammatory responses associated with strokes and their positive and negative effects on the brain in stroke patients, as well as the general dynamics of microglial effects on neural stem cells, both during stroke and in the recovery stage. In the future, we will expand this work to study the mechanisms that improve, stimulate, and generate the neural stem cells in the early stage, and where that information could contribute to understanding the effects of therapeutic strategies. Additionally, it could be interesting to incorporate the dynamics of anti-inflammatory and proinflammatory cytokines from microglia and the cytokines of endogenous neural stem cells into the SMNR model, in order to describe the interaction processes of the different types of cytokines in ischemic stroke.

Author Contributions: Conceptualization, A.J.A.; funding acquisition, A.S.R.; methodology, A.J.A.; Project administration, A.S.R.; supervision, A.S.R. and I.H.; validation, A.S.R.; writing—original draft, A.J.A.; writing—review and editing, I.H. All authors have read and agreed to the published version of the manuscript.

Funding: This research is funded by a grant from Universiti Kebangsaan Malaysia.

Acknowledgments: This study was supported by a research grant from the Universiti Kebangsaan Malaysia with code GUP-2017-112.

Conflicts of Interest: The authors declare no conflict of interest.

References

1. Di Russo, C. A mathematical model of inflammation during ischemic stroke. *Math. Model. Med.* **2010**, *30*, 15–33. [CrossRef]
2. Guruswamy, R.; ElAli, A. Complex roles of microglial cells in ischemic stroke pathobiology: New insights and future directions. *Int. J. Mol. Sci.* **2017**, *18*, 496. [CrossRef]
3. Tobin, M.K.; Bonds, J.A.; Minshall, R.D.; Pelligrino, D.A.; Testai, F.D.; Lazarov, O. Neurogenesis and inflammation after ischemic stroke: What is known and where we go from here. *J. Cereb. Blood Flow Metab.* **2014**, *10*, 1573–1584. [CrossRef] [PubMed]
4. Radak, D.; Katsiki, N.; Resanovic, I.; Jovanovic, A.; Sudar-Milovanovic, E.; Zafirovic, S.; Mousad, S.A.; Isenovic, E.R. Apoptosis and acute brain ischemia in ischemic stroke. *Curr. Vasc. Pharmacol.* **2017**, *15*, 115–122. [CrossRef] [PubMed]
5. Anttila, J.E.; Whitaker, K.W.; Wires, E.S.; Harvey, B.K.; Airavaara, M. Role of microglia in ischemic focal stroke and recovery: Focus on toll-like receptors. *Neuro-Psychopharmacol. Biol. Psychiatry* **2016**, *79*, 3–14. [CrossRef] [PubMed]

6. Stonesifer, C.; Corey, S.; Ghanekar, S.; Diamandis, Z.; Acosta, S.A.; Borlongan, C.V. Stem cell therapy for abrogating stroke-induced neuroinflammation and relevant secondary cell death mechanisms. *Prog. Neurobiol. Prog. Neurobiol.* **2017**, *158*, 94–131. [CrossRef]
7. Banjara, M.; Ghosh, C. Sterile neuroinflammation and strategies for therapeutic intervention. *Int. J. Inflamm.* **2017**, *10*, 8385961. [CrossRef]
8. Rajab, N.F.; Musa, S.M.; Ahmad Munawar, M.; Leong, L.M.; Heng, K.Y.; Ibrahim, F.W.; Chan, K.M. Antineuroinflammatory effects of hibiscus sabdariffa Linn.(Roselle) on Lipopolysaccharides-induced Microglia and Neuroblastoma Cells. *Malays. J. Health Sci.* **2016**, *14*, 111–117.
9. Rune, L.; Martin, W.; Daniel, G.; Christina, F.; Lasse, D.; Ishar, D.; Bente, F. Microglial cell population dynamics in the injured adult central nervous system. *Brain Res. Rev.* **2005**, *48*, 196–206.
10. Galloway, D.A.; Phillips, A.E.; Owen, D.R.; Moore, C.S. Phagocytosis in the Brain: Homeostasis and Disease. *Front. Immunol.* **2019**, *10*, 790. [CrossRef]
11. Pham-Huy, L.A.; He, H.; Pham-Huy, C. Free radicals, antioxidants in disease and health. *Int. J. Biomed. Sci. IJBS* **2008**, *4*, 89–96. [PubMed]
12. Di Meo, S.; Reed, T.T.; Venditti, P.; Victor, V.M. Role of ROS and RNS Sources in Physiological and Pathological Conditions. *Oxid. Med. Cell. Longev.* **2016**, *2016*, 1245049. [CrossRef]
13. Fresta, C.G.; Chakraborty, A.; Wijesinghe, M.B.; Amorini, A.M.; Lazzarino, G.; Lazzarino, G.; Tavazzi, B.; Lunte, S.M.; Caraci, F.; Dhar, G.; et al. Non-toxic engineered carbon nanodiamond concentrations induce oxidative/nitrosative stress, imbalance of energy metabolism, and mitochondrial dysfunction in microglial and alveolar basal epithelial cells. *Cell Death Dis.* **2018**, *9*, 245. [CrossRef] [PubMed]
14. Vay, S.U.; Flitsch, L.J.; Rabenstein, M.; Rogall, R.; Blaschke, S.; Kleinhaus, J.; Bach, A.; Fink, G.R.; Schroeter, M.; Rueger, A.M. The plasticity of primary microglia and their multifaceted effects on endogenous neural stem cells in vitro and in vivo. *J. Neuroinflamm.* **2018**, *15*, 226. [CrossRef]
15. Hui-Yin, Y.; Ahmad, N.; Azmi, N.; Makmor-Bakry, M. Curcumin: The molecular mechanisms of action in inflammation and cell death during kainate-induced epileptogenesis. *Indian J. Pharm. Edu. Res.* **2018**, *52*, 32–41. [CrossRef]
16. Hake, I.; Schönenberger, S.; Neumann, J.; Franke, K.; Paulsen-Merker, K.; Reymann, K.; Ismail, G.; Bin Din, L.; Said, I.M.; Latiff, A.; et al. Neuroprotection and enhanced neurogenesis by extract from the tropical plant Knema laurina after inflammatory damage in living brain tissue. *J. Neuroimmunol.* **2009**, *206*, 91–99. [CrossRef]
17. Cherry, J.D.; Olschowka, J.A.; O'Banion, M.K. Neuroinflammation and M_2 microglia: The good, the bad, and the inflamed. *J. Neuroinflamm.* **2014**, *11*, 98. [CrossRef]
18. Ma, Y.; Wang, J.; Wang, Y.; Yang, G.Y. The biphasic function of microglia in ischemic stroke. *Prog. Neurobiol.* **2016**, *157*, 247–272. [CrossRef]
19. Choi, J.Y.; Kim, J.Y.; Kim, J.Y.; Park, J.; Lee, W.T.; Lee, J.E. M_2 phenotype microglia-derived cytokine stimulates proliferation and neuronal differentiation of endogenous stem cells in ischemic brain. *Exp. Neurobiol.* **2017**, *26*, 33–41. [CrossRef]
20. Ahn, S.; Joyner, A. In vivo analysis of quiescent adult neural stem cells responding to Sonic hedgehog. *Nature* **2005**, *437*, 894–897. [CrossRef]
21. Bauer, S . Cytokine control of adult neural stem cells. *Ann. N. Y. Acad. Sci.* **2009**, *1153*, 48–56. [CrossRef] [PubMed]
22. Vieira, M.S.; Santos, A.K.; Vasconcellos, R.; Goulart, V.A.; Parreira, R.C.; Kihara, A.H.; Ulrich, H.; Resende, R.R. Neural stem cell differentiation into mature neurons: Mechanisms of regulation and biotechnological applications. *Biotechnol. Adv.* **2018**, *7*, 1946–1970. [CrossRef] [PubMed]
23. Martínez-Garza, D.M.; Cantú-Rodríguez, O.G.; Jaime-Pérez, J.C.; Gutiérrez-Aguirre, C.H.; Góngora-Rivera, J.F.; Gómez-Almaguer, D. Current state and perspectives of stem cell therapy for stroke. *Med. Univ.* **2016**, *72*, 169–180. [CrossRef]
24. Matarredona, E.R.; Talaveron, R.; Pastor, A.M. Interactions between neural progenitor cells and microglia in the subventricular zone: Physiological implications in the neurogenic niche and after implantation in the injured brain. *Front. Cell. Neurosci.* **2018**, *12*, 268. [CrossRef]
25. Zhang, Z.; Chopp, M. Neural stem cells and ischemic brain. *J. Stroke* **2016**, *18*, 267–272. [CrossRef] [PubMed]

26. Faiz, M.; Sachewsky, N.; Gascón, S.; Bang, K.A.; Morshead, C.M.; Nagy, A. Adult neural stem cells from the subventricular zone give rise to reactive astrocytes in the cortex after stroke. *Cell Stem Cell* **2015**, *17*, 624–634. [CrossRef] [PubMed]
27. Choi, Y.S.; Lee, M.Y.; Sung, K.W.; Jeong, S.W.; Choi, J.S.; Park, H.J.; Kim, O.N.; Lee, S.B.; Kim, S.Y. Regional differences in enhanced neurogenesis in the dentate gyrus of adult rats after transient forebrain ischemia. *Mol. Cells* **2003**, *16*, 232–238.
28. Barkho, B.Z.; Zhao, X. Adult neural stem cells: Response to stroke injury and potential for therapeutic applications. *Curr. Stem Cell Res. Ther.* **2011**, *6*, 327–338. [CrossRef]
29. Ariffin, S.H.Z.; Wahab, R.M.A.; Ismail, I.; Mahadi, N.M.; Ariffin, Z.Z. Stem cells, cytokines and their receptors. *Asia-Pac. J. Mol. Biol. Biotechnol.* **2005**, *13*, 1–13.
30. Lira-Diaz, E; Gonzalez-Perez, O. Emerging roles of microglia cells in the regulation of adult neural stem cells. *Neuroimmunol. Neuroinflamm.* **2008**, *3*, 204–206.
31. Boese, A.C.; Le, Q.S.E.; Pham, D.; Hamblin, M.H.; Lee, J.P. Neural stem cell therapy for subacute and chronic ischemic stroke. *Stem Cell Res. Ther.* **2018**, *9*. [CrossRef] [PubMed]
32. Reynolds, A.; Rubin, J.; Clermont, G.; Day, J.; Vodovotz, Y.; Bard Ermentrout, G. A reduced mathematical model of the acute inflammatory response: I. derivation of model and analysis of anti-inflammation. *J. Theor. Biol.* **2006**, *242*, 220–236. [CrossRef] [PubMed]
33. Kumar, R.; Clermont, G.; Vodovotz, Y.; Chow, C.C. The dynamics of acute inflammation. *J. Theor. Biol.* **2004**, *230*, 145–155. [CrossRef] [PubMed]
34. Alharbi, S.; Rambely, A. A dynamic simulation of the immune system response to inhibit and eliminate abnormal cells. *Symmetry* **2019**, *11*, 572. [CrossRef]
35. Hao, W.; Friedman, A. Mathematical model on Alzheimer'sdisease. *BMC Syst. Biol.* **2016**, *10*, 108. [CrossRef]
36. Nakata, Y.; Getto, P.; Marciniak-Czochra, A.; Alarcón, T. Stability analysis of multi-compartment models for cell production systems. *J. Biol. Dyn.* **2012**, *6*, 2–18. [CrossRef]
37. Ziebell, F.; Martin-Villalba, A.; Marciniak-Czochra, A. Mathematical modelling of adult hippocampal neurogenesis: Effects of altered stem cell dynamics on cell counts and bromodeoxyuridine-labelled cells. *J. R. Soc. Interface* **2014**, *11*, 20140144. [CrossRef]
38. Cacao, E.; Cucinotta, F.A. Modeling impaired hippocampal neurogenesis after radiation exposure. *Radiat. Res* **2016**, *185*, 319–331. [CrossRef]
39. Huang, L.; Zhang, L. Neural stem cell therapies and hypoxic-ischemic brain injury. *Prog. Neurobiol.* **2018**, *173*. [CrossRef]
40. Leah, E.V.; Prerna, R.R.; Raj, G.K.; Amy, K.; Jonathan, E.R. A mathematical model of neuroinflammation in severe clinical traumatic brain injury. *J. Neuroinflamm.* **2018**, *15*, 345. [CrossRef]
41. Zhao, L.; Zhang, J. *Cellular Therapy for Stroke and CNS Injuries*; Springer Series in Translational Stroke Research; Springer: Berlin, Germany, 2015; pp. 34–39.
42. Loane, D.J.; Kumar, A. Microglia in the TBI brain: The good, the bad, and the dysregulated. *Exp. Neurol.* **2016**, *275*, 316–327. [CrossRef] [PubMed]
43. Kim, J.Y.; Kim, N.; Yenari, M.A. Mechanisms and potential therapeutic applications of microglial activation after brain injury. *CNS Neurosci. Ther.* **2015**, *21*, 309–319. [CrossRef] [PubMed]
44. Saha Ray, S.; Sahoo, S. *Generalized Fractional Order Differential Equations Arising in Physical Models*; CRC Press: Boca Raton, FL, USA, 2018.
45. Gantmacher, F.R. *The Theory of Matrices*; American Mathematical Society: Providence, RI, USA, 1959; Volume 2.

© 2020 by the authors. Licensee MDPI, Basel, Switzerland. This article is an open access article distributed under the terms and conditions of the Creative Commons Attribution (CC BY) license (http://creativecommons.org/licenses/by/4.0/).

Article

Construction of Cubic Timmer Triangular Patches and its Application in Scattered Data Interpolation

Fatin Amani Mohd Ali [1], Samsul Ariffin Abdul Karim [2,*], Azizan Saaban [3], Mohammad Khatim Hasan [4], Abdul Ghaffar [5], Kottakkaran Sooppy Nisar [6,*] and Dumitru Baleanu [7,8]

[1] Fundamental and Applied Sciences Department, Universiti Teknologi PETRONAS, Bandar Seri Iskandar, Seri Iskandar 32610, Perak Darul Ridzuan, Malaysia; fatin_18001405@utp.edu.my
[2] Fundamental and Applied Sciences Department and Centre for Smart Grid Energy Research (CSMER), Institute of Autonomous System, Universiti Teknologi PETRONAS, Bandar Seri Iskandar, Seri Iskandar 32610, Perak Darul Ridzuan, Malaysia
[3] School of Quantitative Sciences, UUMCAS, Universiti Utara Malaysia, Sintok, Kedah 06010, Malaysia; azizan.s@uum.edu.my
[4] Centre for Artificial Intelligence Technology, Faculty of Information Science and Technology, Universiti Kebangsaan Malaysia, UKM Bangi, Selangor 43600, Malaysia; mkh@ukm.edu.my
[5] Department of Mathematical Sciences, BUITEMS, Quetta 87300, Pakistan; abdulghaffar.jaffar@gmail.com
[6] Department of Mathematics, College of Arts and Sciences, Prince Sattam bin Abdulaziz University, Wadi Aldawaser 11991, Saudi Arabia
[7] Department of Mathematics, Cankaya University, 06530 Ankara, Turkey; dumitru@cankaya.edu.tr
[8] Institute of Space Sciences, 077125 Magurele, Romania
* Correspondence: samsul_ariffin@utp.edu.my (S.A.A.K.); ksnisar1@gmail.com or n.sooppy@psau.edu.sa (K.S.N.)

Received: 14 November 2019; Accepted: 26 November 2019; Published: 22 January 2020

Abstract: This paper discusses scattered data interpolation by using cubic Timmer triangular patches. In order to achieve C^1 continuity everywhere, we impose a rational corrected scheme that results from convex combination between three local schemes. The final interpolant has the form quintic numerator and quadratic denominator. We test the scheme by considering the established dataset as well as visualizing the rainfall data and digital elevation in Malaysia. We compare the performance between the proposed scheme and some well-known schemes. Numerical and graphical results are presented by using Mathematica and MATLAB. From all numerical results, the proposed scheme is better in terms of smaller root mean square error (RMSE) and higher coefficient of determination (R^2). The higher R^2 value indicates that the proposed scheme can reconstruct the surface with excellent fit that is in line with the standard set by Renka and Brown's validation.

Keywords: scattered data interpolation; cubic timmer triangular patches; cubic ball triangular patches; cubic Bezier triangular patches; convex combination

1. Introduction

Many computer graphics and vision problems involve scattered data interpolation (SDI). SDI methods aims to build a smooth function from a set of data, which consist of functional values corresponding to points, which do not obey any structure or order between their relative locations. These methods have a wide range of uses in surface reconstruction, visualization, image restoration, computer graphics, surface deformation, image processing, engineering, and technology, etc.

Most researchers have investigated surface interpolation based on triangulations of scattered data and there are several scattered data fitting techniques, such as the Delaunay triangulation method [1], radial basis function (RBF) [2], and moving least square (MLS) [3]. Very recently, new techniques for interpolating scattered data have been developed [1,4,5], which can be implemented in fast algorithms [6].

The most popular method that is usually used to generate the surface of scattered from the data points is Delaunay triangulation method [7]. It is a very famous method, applied to produce the triangle meshes, where vertices of the triangle are made up of the sample data points.

The property of shape preserving interpolation is an important technique usually applied in curve and surface modeling. Several research papers [2,7–13] have been published on shape preservation in the last couple of years. Ibraheem et al. [14] proposed a scheme that is suitable for surface reconstruction and deformation. The objective of the scheme is to develop a local positivity preserving when the data points are used.

A new big data infrastructure for the management of cultural items was proposed by Su et al. [15]. It is a multilayer architecture to create new applications based on the modules that were offered by APIs. Streams of data from social networks are mostly captured by this module to handle and update the information. They tested their system by created an application of the Android devices called the Smart Search Museum. This application will access the map that has all the museums of a given area in Italy. Nowadays, the demand for information monitoring and recommendation technology is getting bigger, which is suggested by the systems proposed by [1]. They proposed a novel, collaborative, and user-centered approach for big data application.

This study is an extension of the paper discussed by Ali et al. [16]. The main objective of this study is to construct the SDI using Timmer triangular patches, which are used to visualize the energy data i.e., spatial interpolation in visualizing rainfall data. Firstly, we triangulate the domain data using Delaunay triangulation. Next, we specify the derivatives at the data points and assign Timmer ordinate values for each triangular patch. Lastly, we generate the Timmer triangular patches of the surface. The main novelty of this work is that we construct new cubic triangular Timmer patches and apply them for scattered data interpolation. The proposed scheme has higher accuracy as well as requiring smaller CPU times (in seconds) than some existing schemes.

The rest of the paper is organized as follows. The method of the study is discussed in Section 2 including the derivation of cubic Timmer triangular patches with some examples. In Section 3, we discuss the derivation of the sufficient condition for C^1 continuity on all adjacent triangles. Numerical and graphical results including comparison with some established schemes are presented in Section 4. In the final section, some conclusions and recommendations for future studies are made.

2. Cubic Timmer Triangular Patches

Ali et al. [16] introduced a new cubic triangular basis function, which was actually the extension of the univariate cubic Timmer basis proposed by Timmer [17]. The new cubic Timmer triangular patch is different from the cubic Ball and cubic Bezier triangular patches [16].

The cubic Timmer patch is defined as follows [16]:

$$T(u,v,w) = \sum_{i+j+k=3} T^3_{ijk}(u,v,w) t_{i,j,k}. \tag{1}$$

Equation (1) can be written as

$$\begin{aligned} T(u,v,w) = \; & u^2(2u-1)t_{300} + 4u^2vt_{210} + 4u^2wt_{201} \\ & + v^2(2v-1)t_{030} + 4v^2ut_{120} + 4v^2wt_{021} \\ & + w^2(2w-1)t_{003} + 4w^2ut_{102} \\ & + 4w^2vt_{012} + 6uvwt_{111} \end{aligned} \tag{2}$$

where t_{ijk} denotes the control point, while $T^3_{ijk}(u,v,w)$, $i+j+k=3$ are cubic Timmer triangular basis functions defined in [16]. Some properties of the Timmer triangle patch described in Equation (1) are the following:

(a) Partition of unity: The new cubic Timmer triangular basis satisfies:

$$\sum_{i=0}^{3} T_{ijk}^3(u,v,w) = 1.$$

(b) Symmetry: The surfaces generated from two different ordering of its control points will look the same.
(c) Positivity: In each of the cubic Timmer triangular basis functions, the positivity or nonnegativity behavior $T_{ijk}^3(u,v,w) \geq 0$ is fulfilled, except for the following: $T_{300}^3(u,v,w) \leq 0$ when $\frac{1}{2} \leq u \leq 1$ and both $T_{201}^3(u,v,w) \leq 0$ and $T_{210}^3(u,v,w) \leq 0$ when $0 \leq u \leq \frac{1}{2}$.
(d) Convex hull: The Timmer triangular patches may not lie within the convex hull of the control polygon. If the positivity property is fulfilled for the Timmer triangular patches as discussed in (c), it will satisfy the convex hull property.

Figures 1–3 show the cubic Timmer triangular basis functions, the Timmer ordinates for the cubic Timmer triangular patch, and the cubic Timmer triangular bases, respectively.

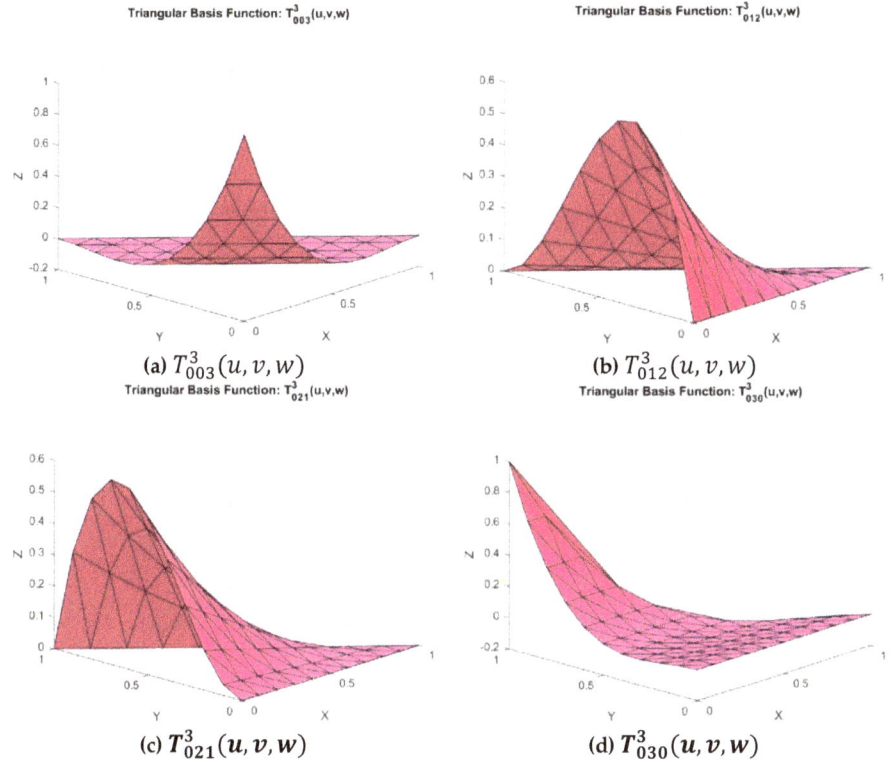

Figure 1. Cont.

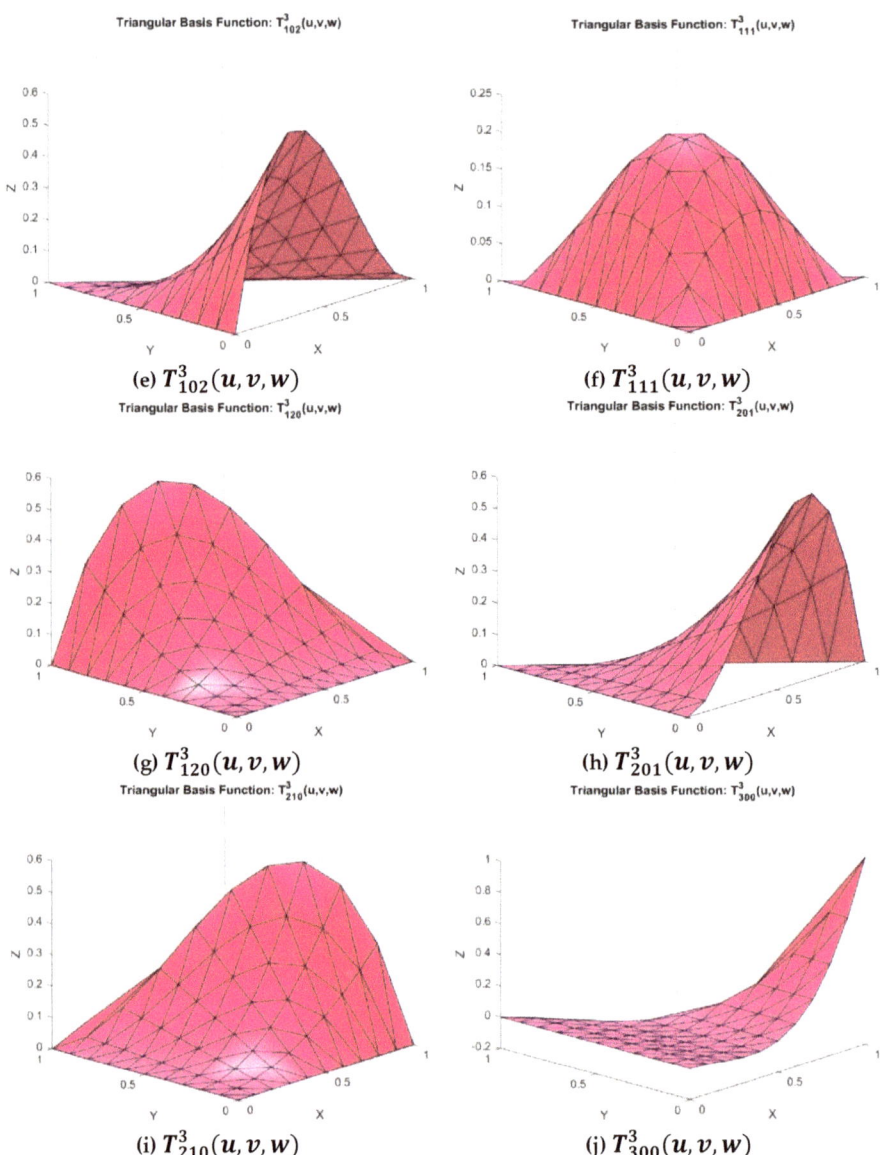

Figure 1. Some cubic Timmer triangular basis functions.

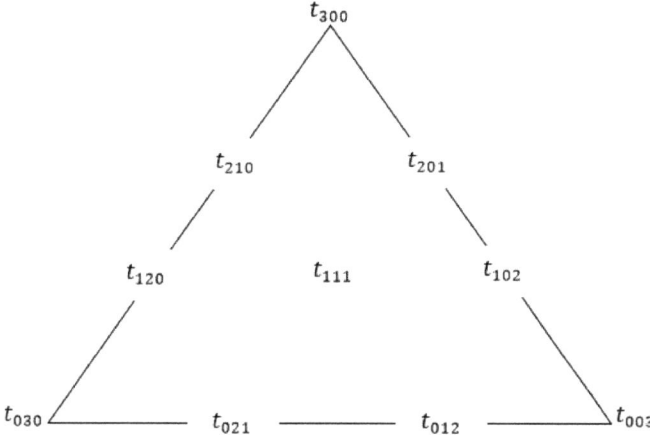

Figure 2. Control nets for cubic Timmer triangular patch.

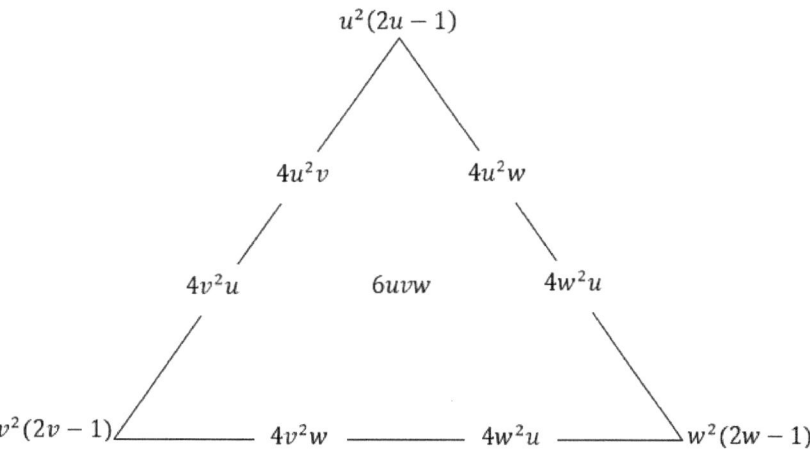

Figure 3. Cubic Timmer triangular bases.

Then, one set of control points is used to construct the surface of cubic Timmer triangular patch. Table 1 shows the control point used to construct the surfaces in Figure 4.

Table 1. Control points.

x	−4.0	−4.0	−4.0	−4.0	−1.5	1.0	3.5	1.0	−1.5	−1.5
y	3.5	1.0	−1.5	−4.0	−4.0	−4.0	−4.0	−1.5	−1.5	−1.5
z	20.25	9.00	1.0.25	24.00	10.25	9.00	20.25	4.75	4.75	3.50

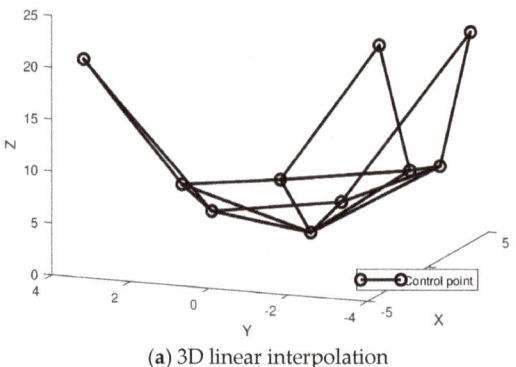

(a) 3D linear interpolation

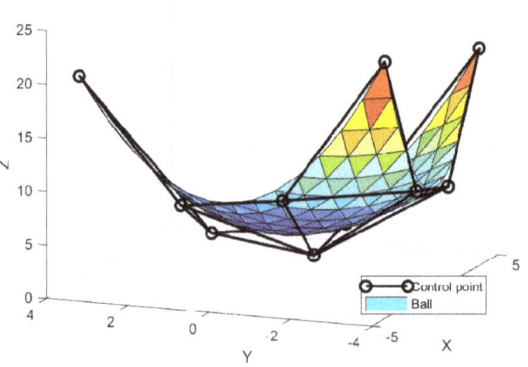

(b) Cubic Timmer triangular patch

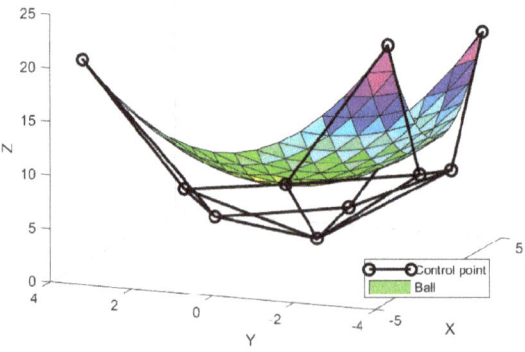

(c) Cubic Bèzier triangular patch

Figure 4. Cont.

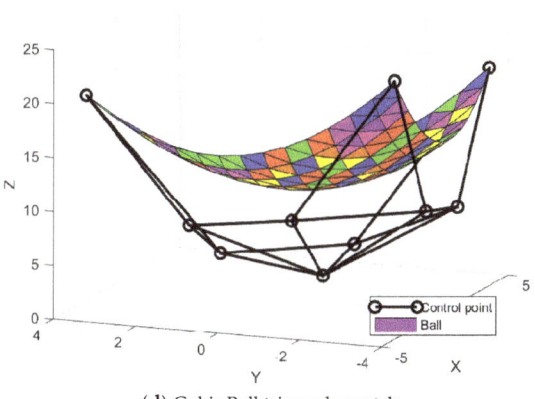

(**d**) Cubic Ball triangular patch

Figure 4. Surface interpolation.

It is observed from Figure 4 that cubic Timmer triangular patches lie in close vicinity of the control polygon as compared to cubic Bèzier and Ball triangular patches, respectively. The cubic Bèzier and Ball triangular patches satisfy the convex hull property while the cubic Timmer triangular, in general, does not obey the convex hull property. However, cubic Timmer triangular patches are close to the control polyhedron.

3. Derivation of Sufficient Condition for C^1 Continuity on Adjacent Triangles

In this section, we show in detail the derivation of the sufficient condition that have been considered in Ali et al. [2]. The derivatives of T with respect to the direction $s = (s_x, s_y, s_z) = s_x V_1 + s_y V_2 + s_z V_3, s_x + s_y + s_z = 0$ are given by:

$$D_s T(u,v,w) = s_x \frac{\partial T}{\partial u} + s_y \frac{\partial T}{\partial v} + s_z \frac{\partial T}{\partial w}. \qquad (3)$$

From Equation (2), it can be shown that

$$\left. \begin{array}{l} \frac{\partial T}{\partial u} = 4v^2 t_{120} + 4w^2 t_{102} + 6vw t_{111}^1 \\ \frac{\partial T}{\partial v} = (6v^2 - 2v)t_{030} + 8vw t_{021} + 4w^2 t_{012} \\ \frac{\partial T}{\partial w} = (6w^2 - 2w)t_{003} + 4v^2 t_{021} + 8vw t_{012} \end{array} \right\}. \qquad (4)$$

Local Scheme

Consider a triangle with vertices W_1, W_2, W_3, barycentric coordinates u, v, w, and edges $e_1 = (0, -1, 1), e_2 = (1, 0, -1), e_3 = (0, 0, 1)$. Any point on the triangle can be expressed as

$$W = uW_1 + vW_2 + wW_3, \; u + v + w = 1. \qquad (5)$$

We use the following two methods of convex combination among the three local schemes as described below:

$$T(u,v,w) = \frac{vw T_1 + uw T_2 + uv T_3}{vw + uw + uv} \qquad (6)$$

where the local scheme $T_i, i = 1, 2, 3$ is obtained by replacing the inner ordinate with t_{111}^i, $i = 1, 2, 3$. The derivations of all three local schemes are described in the following paragraphs.

Let n_1, n_2, n_3 be the inward normal direction to the line segment W_2W_3, W_3W_1, W_1W_2 as shown in Figure 5, where

$$n_1 = -e_3 + \frac{e_3.e_1}{|e_1|^2}e_1, n_2 = -e_1 + \frac{e_1.e_2}{|e_2|^2}e_2, n_1 = -e_2 + \frac{e_2.e_3}{|e_3|^2}e_3 \qquad (7)$$

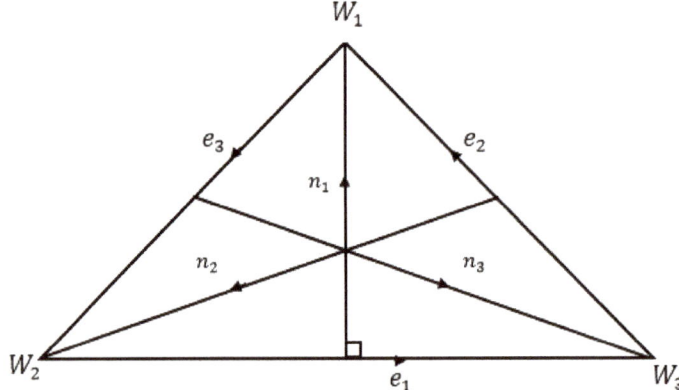

Figure 5. Inward normal direction to the edges of triangle.

The normal derivatives of local scheme T_1, T_2, T_3 are defined by

$$\left.\begin{array}{rl} D_{n_1}T_1 = & (4t_{120} - 2t_{021} - 3t_{030})v^2 + (4t_{102} - 2t_{012} - 3t_{003})w^2 \\ & +2\left(3t_{111}^1 - 2t_{021} - 2t_{012}\right)vw + vt_{030} + wt_{003} \\ D_{n_2}T_2 = & (4t_{120} - 2t_{021} - 3t_{030})u^2 + (4t_{102} - 2t_{012} - 3t_{003})w^2 \\ & +2\left(3t_{111}^2 - 2t_{021} - 2t_{012}\right)vw + vt_{030} + wt_{003} \\ D_{n_3}T_3 = & (4t_{120} - 2t_{021} - 3t_{030})u^2 + (4t_{102} - 2t_{012} - 3t_{003})v^2 \\ & +2\left(3t_{111}^2 - 2t_{021} - 2t_{012}\right)uv + ut_{030} + vt_{003} \end{array}\right\}. \qquad (8)$$

The boundary ordinates are given as follows [2]:

$$t_{210} = t_{300} + \frac{1}{4}\left[(x_2 - x_1)T_x(W_1) + (y_2 - y_1)T_y(W_1)\right],$$

$$t_{201} = t_{300} + \frac{1}{4}\left[(x_1 - x_3)T_x(W_1) + (y_1 - y_3)T_y(W_1)\right],$$

$$t_{012} = t_{030} + \frac{1}{4}\left[(x_3 - x_2)T_x(W_2) + (y_3 - y_2)T_y(W_2)\right],$$

$$t_{120} = t_{030} + \frac{1}{4}\left[(x_2 - x_1)T_x(W_2) + (y_2 - y_1)T_y(W_2)\right],$$

$$t_{102} = t_{003} + \frac{1}{4}\left[(x_1 - x_3)T_x(W_3) + (y_1 - y_3)T_y(W_3)\right],$$

and

$$t_{012} = t_{003} + \frac{1}{4}\left[(x_3 - x_2)T_x(W_3) + (y_3 - y_2)T_y(W_3)\right].$$

where the first partial derivatives $T_x(W_1)$, $T_y(W_1)$ are estimated by using Goodman et al. [18] method. The inner ordinates i.e., t_{111}^i, $i = 1, 2, 3$ are obtained by using two different methods i.e., Goodman and Said [8] and Foley and Opitz [19]. The following paragraphs describes both methods.

The inner ordinates by using Goodman and Said [8] method are given as follows [2]:

$$a^1_{111} = \frac{2}{3}(a_{120} + a_{102}) + \frac{1}{3}(a_{021} + a_{012}) - \frac{1}{2}(a_{030} + a_{003}),$$

$$a^2_{111} = \frac{2}{3}(a_{210} + a_{012}) + \frac{1}{3}(a_{201} + a_{102}) - \frac{1}{2}(a_{300} + a_{003}),$$

and

$$a^3_{111} = \frac{2}{3}(a_{201} + a_{021}) + \frac{1}{3}(a_{210} + a_{120}) - \frac{1}{2}(a_{300} + a_{030}).$$

Foley and Opitz's [19] method is as follows. Assume two adjacent triangles, L_1, L_2 as shown in Figure 6 with vertices E_i and F_i with e_1 as a common edge (see Figure 6).

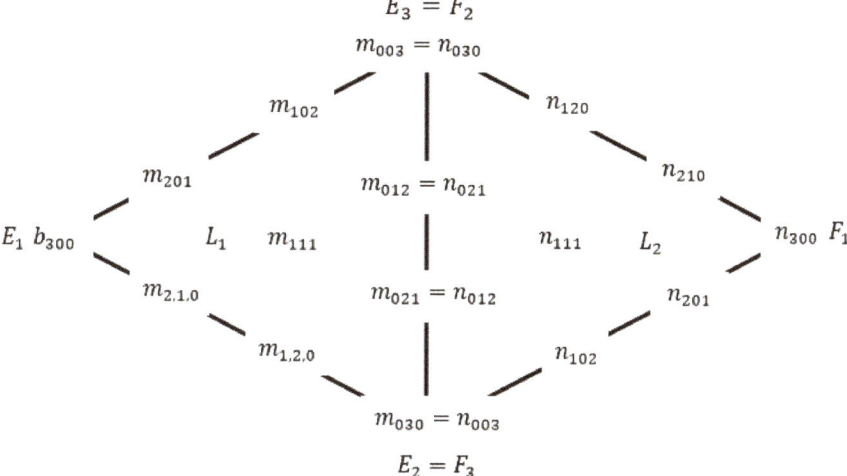

Figure 6. Two adjacent cubic triangular patches for Foley and Opitz [19].

The following equations must be satisfied to produce C^1 continuity along the common edge.

$$n_{201} = r^2 m_{210} + 2stm_{021} + 2rsm_{120} + s^2 m_{030} + 2rtm^1_{111} + t^2 m_{012}, \quad (9)$$

$$n_{210} = r^2 m_{201} + 2stm_{012} + 2rtm_{102} + s^2 m_{021} + 2rsm^1_{111} + t^2 m_{003}, \quad (10)$$

$$m_{201} = u^2 n_{201} + 2vwn_{012} + 2uwn_{102} + v^2 n_{021} + 2uvn^1_{111} + w^2 n_{003}, \quad (11)$$

$$m_{201} = u^2 n_{210} + 2vwn_{021} + 2uvn_{120} + v^2 n_{030} + 2uwn^1_{111} + w^2 n_{012}, \quad (12)$$

where $F_1 = rE_1 + sE_2 + tE_3$ and $E_1 = uF_1 + vF_2 + wF_3$. Equations (9) and (10) will be added together to obtain the inner ordinate m^1_{111}. Then, solve for m^1_{111}:

$$m^1_{111} = \frac{1}{2u(v+w)}(n_{201} + n_{210}) - u^2(m_{210} + m_{201}) - v^2(m_{030} + m_{021}) \\ - w^2(m_{012} + m_{003}) - 2vw(m_{021} + m_{012}) - uvm_{120} - 2uwm_{102}. \quad (13)$$

This calculation is similar to obtain the inner ordinate n^1_{111}. Adding the Equations (11) and (12) and the value of n^1_{111} is given as:

$$n^1_{111} = \frac{1}{2r(s+t)}(m_{201} + m_{210}) - r^2(n_{210} + n_{201}) - s^2(n_{030} + n_{021}) - t^2(n_{012} + n_{003}) \\ - 2st(n_{021} + n_{012}) - rsn_{120} - 2rtn_{102}. \quad (14)$$

To produce the final interpolant, two methods of convex combination mentioned in Equation (6) will be used. The final scheme of scattered data interpolation using cubic Timmer triangular patch can be expressed as Theorem 3.

Theorem 3: *The final interpolating surface T on each triangle can be expressed as:*

$$T(u,v,w) = \alpha_1 T_1(u,v,w) + \alpha_2 T_2(u,v,w) + \alpha_3 T_3(u,v,w)$$

where

$$\alpha_1 = \frac{vw}{vw+uv+uw}, \alpha_2 = \frac{uw}{vw+uv+uw}, \alpha_3 = \frac{uv}{vw+uv+uw}$$

Equivalently

$$T(u,v,w) = \sum_{\substack{i+j+k=3,\\ i\neq 1,\ j\neq 1,\ k\neq 1}} t_{ijk} T(u,v,w) + 6uvw\left(\alpha_1 t^1_{111} + \beta_2 t^2_{111} + \gamma_3 t^3_{111}\right). \quad (15)$$

Another version of convex combination scheme is presented in Goodman and Said [8] and is given as follows:

$$\alpha_1 = \frac{v^2 w^2}{v^2 w^2 + u^2 v^2 + u^2 w^2}, \alpha_2 = \frac{u^2 w^2}{v^2 w^2 + u^2 v^2 + u^2 w^2}, \alpha_3 = \frac{u^2 v^2}{v^2 w^2 + u^2 v^2 + u^2 w^2}. \quad (16)$$

For the purpose of numerical comparison later, we denoted convex combination used in Equation (6) as Choice 1; meanwhile, Goodman and Said [8] is Choice 2.

Theorem 4: *The rational corrected interpolant defined by (8) is in the form quantic numerator and quadratic denominator.*

Proof. From Equation (8), the resulting interpolant is degree 7 i.e., degree five in numerator and degree two in denominator. □

The following Algorithm 1 can be used to construct surface form scattered date.

Algorithm 1. Construction scattered data interpolation

Input: Data points

1. Triangulate the domain by using Delaunay triangulation.
2. Specify the derivatives at the data points using [18] then assign Timmer ordinates values for each triangular patch.
3. Generate the triangular patches of the surfaces by using cubic Timmer triangular patches.
4. Calculate CPU time (in seconds), RMSE, maximum error and R^2.

Output: Surface reconstruction
Steps 1–4 are repeated for different test function.

4. Results and Discussions

To test the capability of the proposed scattered data interpolation scheme, we use six well-known test functions $F_1(x,y), F_2(x,y), F_3(x,y), F_4(x,y), F_5(x,y)$, and $F_6(x,y)$ as shown in Figure 7 [20]. All numerical simulation and graphical visualization are done by using MATLAB R2019a version on Intel® Core i5–6200U 2.3GHz with Turbo Boost up to 2.8GHz.

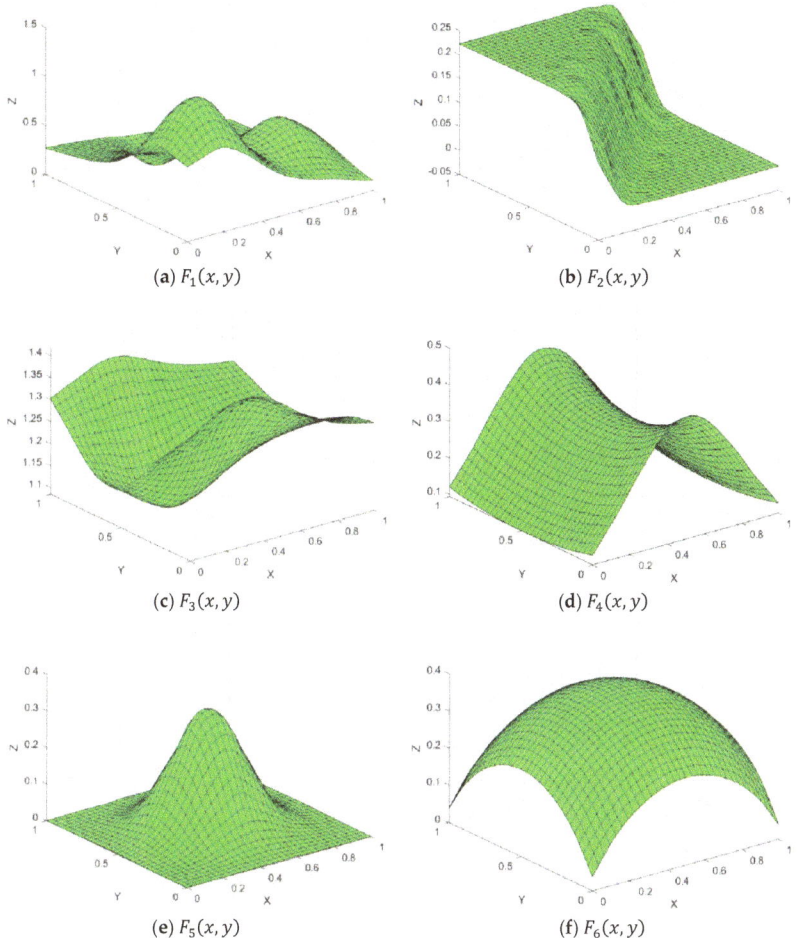

Figure 7. Test functions.

1. Franke's exponential function

$$F_1(x,y) = 0.75e^{-(\frac{(9x-2)^2+(9y-2)^2}{4})} + 0.75e^{-(\frac{(9x+1)^2}{49}+\frac{9y+1}{10})} + 0.50e^{-(\frac{(9x-7)^2+(9y-3)^2}{4})} - 0.20e^{-((9x-4)^2+(9y-7)^2)};$$

2. Cliff function

$$F_2(x,y) = \frac{\tanh(9y-9x)+1}{9};$$

3. Saddle function

$$F_3(x,y) = \frac{1.25+\cos(4.5y)}{6+6(3x-1)^2};$$

4. Gentle function

$$F_4(x,y) = \frac{exp\left(-\left(\frac{81}{16}\right)\left((x-0.5)^2 + (y-0.5)^2\right)\right)}{3};$$

5. Steep function

$$F_5(x,y) = \frac{exp\left(-\left(\frac{81}{4}\right)\left((x-0.5)^2 + (y-0.5)^2\right)\right)}{3};$$

6. Sphere function

$$F_6(x,y) = \frac{1}{9}\sqrt{64 - 81\left((x-0.5)^2 + (y-0.5)^2\right)} - 0.5.$$

We use three datasets i.e., 36, 65, and 100 number of points, as shown in Tables 2–4.

Table 2. Thirty-six datasets.

X	Y	$F_1(x,y)$	$F_2(x,y)$	$F_3(x,y)$	$F_4(x,y)$	$F_5(x,y)$	$F_6(x,y)$
0.0000	0.0000	0.7664	0.1111	1.3333	0.1207	1.34E-05	0.0386
0.5000	0.0000	0.4349	2.74E-05	1.3833	0.4280	0.0021	0.2349
1.0000	0.0000	0.1076	3.38E-09	1.2833	0.1207	1.34E-05	0.0386
0.1500	0.1500	1.1370	0.1111	1.3382	0.2027	0.0023	0.2383
0.7000	0.1500	0.4304	1.11E-05	1.3020	0.3077	0.0124	0.2922
0.5000	0.2000	0.5345	0.0010	1.3128	0.3647	0.0539	0.3367
0.2500	0.3000	1.0726	0.1580	1.2423	0.2528	0.0418	0.3292
0.4000	0.3000	0.7134	0.0315	1.2421	0.3298	0.1211	0.3603
0.7500	0.4000	0.5903	0.0004	1.2139	0.2454	0.0768	0.3471
0.8500	0.2500	0.5088	4.53E-06	1.2607	0.1908	0.0079	0.2779
0.5500	0.4500	0.3823	0.0315	1.1613	0.3300	0.3012	0.3861
0.0000	0.5000	0.4818	0.2222	1.1747	0.0940	0.0021	0.2349
0.2000	0.4500	0.6458	0.2198	1.1412	0.2119	0.0512	0.3352
0.4500	0.5500	0.2946	0.1907	1.1037	0.3300	0.3012	0.3861
0.6000	0.6500	0.1920	0.1580	1.1552	0.3241	0.1726	0.3704
0.2500	0.7000	0.2930	0.2222	1.1240	0.2528	0.0418	0.3292
0.4000	0.8000	0.0515	0.2221	1.1887	0.3467	0.0440	0.3307
0.6500	0.7500	0.1372	0.1907	1.1961	0.3166	0.0596	0.3397
0.8000	0.8500	0.0823	0.1580	1.2431	0.2389	0.0045	0.2600
0.8500	0.6500	0.1412	0.0059	1.2043	0.1834	0.0177	0.3032
1.0000	0.5000	0.1610	2.74E-05	1.2199	0.0940	0.0021	0.2349
1.0000	1.0000	0.0359	0.1111	1.2712	0.1207	1.34E-05	0.0386
0.5000	1.0000	0.1460	0.2222	1.3346	0.4280	0.0021	0.2349
0.1000	0.8500	0.2935	0.2222	1.2363	0.1676	0.0011	0.2125
0.0000	1.0000	0.2703	0.2222	1.3029	0.1207	1.34E-05	0.0386
0.2500	0.0000	0.8189	0.0024	1.4069	0.3119	0.0006	0.1911

Table 2. *Cont.*

X	Y	$F_1(x,y)$	$F_2(x,y)$	$F_3(x,y)$	$F_4(x,y)$	$F_5(x,y)$	$F_6(x,y)$
0.7500	0.0000	0.2521	3.05E-07	1.3150	0.3119	0.0006	0.1911
0.2500	1.0000	0.2222	0.2222	1.3496	0.3119	0.0006	0.1911
0.0000	0.2500	0.8026	0.2198	1.2683	0.1001	0.0006	0.1911
0.7500	1.0000	0.0810	0.2198	1.2913	0.3119	0.0006	0.1911
0.0000	0.7500	0.3395	0.2222	1.1987	0.1001	0.0006	0.1911
1.0000	0.2500	0.2302	3.05E-07	1.2573	0.1001	0.0006	0.1911
1.0000	0.7500	0.0504	0.0024	1.2295	0.1001	0.0006	0.1911
0.1900	0.1900	1.2118	0.1111	1.3229	0.2256	0.0068	0.2733
0.3200	0.7500	0.2029	0.2221	1.1477	0.3012	0.0488	0.3338
0.7900	0.4600	0.4777	0.0006	1.2041	0.2181	0.0588	0.3393

Table 3. Sixty-five datasets.

X	Y	$F_1(x,y)$	$F_2(x,y)$	$F_3(x,y)$	$F_4(x,y)$	$F_5(x,y)$	$F_6(x,y)$
0.0500	0.4500	0.5775	0.0024	1.1767	0.3119	0.0052	0.2649
0.0000	0.5000	0.4818	3.05E-07	1.1747	0.3119	0.0021	0.2349
0.0000	1.0000	0.2703	3.05E-07	1.3029	0.1001	1.34E-05	0.0386
0.0000	0.0000	0.7664	0.0024	1.3333	0.1001	1.34E-05	0.0386
0.1000	0.1500	1.0495	0.2198	1.3271	0.3119	0.0011	0.2125
0.1000	0.7500	0.3229	0.2222	1.1812	0.3119	0.0037	0.2534
0.1500	0.3000	1.0600	0.2222	1.2437	0.1001	0.0124	0.2922
0.2000	0.1000	1.0774	0.2198	1.3732	0.1001	0.0021	0.2349
0.2500	0.2000	1.1892	0.2010	1.3239	0.2100	0.0152	0.2985
0.3000	0.3500	0.8562	2.89E-06	1.1982	0.0955	0.0940	0.3530
0.3500	0.8500	0.1588	0.0212	1.2297	0.1082	0.0177	0.3032
0.5000	0.0000	0.4349	0.2220	1.3833	0.3955	0.0021	0.2349
0.5000	1.0000	0.1460	0.2220	1.3346	0.0955	0.0021	0.2349
0.5500	0.9500	0.1329	0.2010	1.2975	0.1082	0.0052	0.2649
0.2500	0.0000	0.8189	0.0004	1.4069	0.3373	0.0006	0.1911
0.7500	0.0000	0.2521	0.1580	1.3150	0.3241	0.0006	0.1911
1.0000	0.2500	0.2302	0.2198	1.2573	0.3582	0.0006	0.1911
1.0000	0.7500	0.0504	0.1580	1.2295	0.3096	0.0006	0.1911
0.7500	1.0000	0.0810	2.74E-05	1.2913	0.2979	0.0006	0.1911
0.2500	1.0000	0.2222	0.0642	1.3496	0.2784	0.0006	0.1911
0.0000	0.7500	0.3395	0.2163	1.1987	0.3195	0.0006	0.1911
0.0000	0.2500	0.8026	1.84E-06	1.2683	0.2851	0.0006	0.1911
0.8750	1.0000	0.0553	0.0002	1.2791	0.2484	0.0001	0.1321
1.0000	0.3750	0.2405	0.1907	1.2354	0.2746	0.0015	0.2242
1.0000	0.8750	0.0406	0.0002	1.2504	0.2135	0.0001	0.1321

Table 3. Cont.

X	Y	$F_1(x,y)$	$F_2(x,y)$	$F_3(x,y)$	$F_4(x,y)$	$F_5(x,y)$	$F_6(x,y)$
0.6250	1.0000	0.1124	0.0140	1.3099	0.2162	0.0015	0.2242
0.0000	0.3750	0.6464	4.53E-06	1.2134	0.1908	0.0015	0.2242
0.0000	0.1250	0.8220	1.11E-05	1.3151	0.1517	0.0001	0.1321
0.6000	0.2500	0.5050	0.0315	1.2723	0.1623	0.0768	0.3471
0.6000	0.6500	0.1920	0.0642	1.1552	0.1403	0.1726	0.3704
0.6000	0.8500	0.1196	3.38E-09	1.2376	0.1207	0.0228	0.3109
0.6500	0.7000	0.1585	2.74E-05	1.1796	0.0940	0.0940	0.3530
0.7000	0.2000	0.5070	0.1111	1.2855	0.1207	0.0240	0.3125
0.7000	0.6500	0.1854	0.2221	1.1796	0.4196	0.0940	0.3530
0.7000	0.9000	0.1021	0.2221	1.2611	0.0944	0.0058	0.2682
0.7500	0.1000	0.3475	0.0059	1.3058	0.3494	0.0037	0.2534
0.7500	0.3500	0.6368	0.0212	1.2296	0.3248	0.0596	0.3397
0.7500	0.8500	0.0948	0.2167	1.2421	0.3091	0.0079	0.2779
0.8000	0.4000	0.5729	0.0457	1.2187	0.3341	0.0440	0.3307
0.8000	0.6500	0.1608	0.2222	1.1975	0.2414	0.0342	0.3232
0.8500	0.2500	0.5088	0.2221	1.2607	0.2601	0.0079	0.2779
0.9000	0.3500	0.4588	0.2217	1.2366	0.2132	0.0083	0.2795
0.9000	0.8000	0.0654	3.21E-08	1.2336	0.1082	0.0021	0.2349
0.9500	0.9000	0.0473	3.21E-08	1.2555	0.2100	0.0002	0.1539
1.0000	0.0000	0.1076	0.2207	1.2833	0.3537	1.34E-05	0.0386
1.0000	0.5000	0.1610	0.0005	1.2199	0.1715	0.0021	0.2349
1.0000	1.0000	0.0359	0.0003	1.2712	0.3955	1.34E-05	0.0386
0.5625	1.0000	0.1292	0.0002	1.3218	0.3797	0.0019	0.2323
0.0000	0.4375	0.5564	0.0061	1.1907	0.2690	0.0019	0.2323
0.4250	0.2250	0.7025	0.1634	1.3040	0.3345	0.0643	0.3419
0.5750	0.4500	0.3940	0.2222	1.1673	0.0955	0.2828	0.3843
0.3732	0.5768	0.3115	0.1111	1.0857	0.1207	0.2136	0.3764
0.4475	0.3725	0.5128	0.1111	1.1864	0.1207	0.2268	0.3781
0.2013	0.8592	0.2658	0.1111	1.2395	0.1207	0.0040	0.2562
0.2611	0.7021	0.2857	0.1111	1.1232	0.1207	0.0459	0.3320
0.2024	0.5368	0.4564	0.1111	1.1098	0.1207	0.0540	0.3368
1.0000	0.1250	0.1552	0.1111	1.2760	0.1207	0.0001	0.1321
0.8750	0.0000	0.1796	0.1111	1.2958	0.1207	0.0001	0.1321
0.4750	0.7500	0.0247	0.1111	1.1632	0.1207	0.0928	0.3526
0.8625	0.5250	0.2938	0.1111	1.2049	0.1207	0.0230	0.3112
0.3750	0.0000	0.6331	0.1111	1.4141	0.1207	0.0015	0.2242
0.5273	0.1341	0.4668	0.1111	1.3433	0.1207	0.0218	0.3096
0.7058	0.5073	0.3878	0.1111	1.1818	0.1207	0.1412	0.3647
0.5037	0.5605	0.2694	0.1111	1.1187	0.1207	0.3094	0.3868
0.0000	0.6250	0.3892	0.1111	1.1689	0.1207	0.0015	0.2242

Table 4. One hundred datasets.

X	Y	$F_1(x,y)$	$F_2(x,y)$	$F_3(x,y)$	$F_4(x,y)$	$F_5(x,y)$	$F_6(x,y)$
0.0500	0.4500	0.5775	0.2221	1.1767	0.1199	0.0052	0.2649
0.0000	0.5000	0.4818	0.2222	1.1747	0.0940	0.0021	0.2349
0.0000	1.0000	0.2703	0.2222	1.3029	0.1207	0.0000	0.0386
0.0000	0.0000	0.7664	0.1111	1.3333	0.1207	0.0000	0.0386
0.1000	0.1500	1.0495	0.1580	1.3271	0.1676	0.0011	0.2125
0.1000	0.7500	0.3229	0.2222	1.1812	0.1578	0.0037	0.2534
0.1500	0.3000	1.0600	0.2082	1.2437	0.1866	0.0124	0.2922
0.2000	0.1000	1.0774	0.0315	1.3732	0.2480	0.0021	0.2349
0.2500	0.2000	1.1892	0.0642	1.3239	0.2658	0.0152	0.2985
0.3000	0.3500	0.8562	0.1580	1.1982	0.2784	0.0940	0.3530
0.3500	0.8500	0.1588	0.2222	1.2297	0.3362	0.0177	0.3032
0.5000	0.0000	0.4349	0.0000	1.3833	0.4280	0.0021	0.2349
0.5000	1.0000	0.1460	0.2222	1.3346	0.4280	0.0021	0.2349
0.5500	0.9500	0.1329	0.2221	1.2975	0.4030	0.0052	0.2649
0.6000	0.2500	0.5050	0.0004	1.2723	0.3373	0.0768	0.3471
0.6000	0.6500	0.1920	0.1580	1.1552	0.3241	0.1726	0.3704
0.6000	0.8500	0.1196	0.2198	1.2376	0.3582	0.0228	0.3109
0.6500	0.7000	0.1585	0.1580	1.1796	0.3096	0.0940	0.3530
0.7000	0.2000	0.5070	0.0000	1.2855	0.2979	0.0240	0.3125
0.7000	0.6500	0.1854	0.0642	1.1796	0.2784	0.0940	0.3530
0.7000	0.9000	0.1021	0.2163	1.2611	0.3195	0.0058	0.2682
0.7500	0.1000	0.3475	0.0000	1.3058	0.2851	0.0037	0.2534
0.7500	0.3500	0.6368	0.0002	1.2296	0.2484	0.0596	0.3397
0.7500	0.8500	0.0948	0.1907	1.2421	0.2746	0.0079	0.2779
0.8000	0.4000	0.5729	0.0002	1.2187	0.2135	0.0440	0.3307
0.8000	0.6500	0.1608	0.0140	1.1975	0.2162	0.0342	0.3232
0.8500	0.2500	0.5088	0.0000	1.2607	0.1908	0.0079	0.2779
0.9000	0.3500	0.4588	0.0000	1.2366	0.1517	0.0083	0.2795
0.9000	0.8000	0.0654	0.0315	1.2336	0.1623	0.0021	0.2349
0.9500	0.9000	0.0473	0.0642	1.2555	0.1403	0.0002	0.1539
1.0000	0.0000	0.1076	0.0000	1.2833	0.1207	0.0000	0.0386
0.0000	0.6250	0.3892	0.2222	1.1689	0.0955	0.0015	0.2242
0.6250	0.0000	0.3203	0.0000	1.3444	0.3955	0.0015	0.2242
0.8750	0.0000	0.1796	0.0000	1.2958	0.2100	0.0001	0.1321
0.0000	0.8750	0.3026	0.2222	1.2511	0.1082	0.0001	0.1321
0.4386	0.5114	0.3383	0.1750	1.1093	0.3271	0.3080	0.3867
0.1816	0.5822	0.4015	0.2221	1.1119	0.2009	0.0373	0.3258
0.4250	0.2250	0.7025	0.0059	1.3040	0.3494	0.0643	0.3419
0.4750	0.8500	0.0579	0.2220	1.2328	0.3756	0.0275	0.3167
0.4125	0.6855	0.1477	0.2206	1.1163	0.3319	0.1422	0.3649
0.5993	0.1237	0.4060	0.0000	1.3299	0.3653	0.0155	0.2992

Table 4. *Cont.*

X	Y	$F_1(x,y)$	$F_2(x,y)$	$F_3(x,y)$	$F_4(x,y)$	$F_5(x,y)$	$F_6(x,y)$
1.0000	0.1250	0.1552	0.0000	1.2760	0.1082	0.0001	0.1321
0.1250	1.0000	0.2516	0.2222	1.3261	0.2100	0.0001	0.1321
0.1875	0.8675	0.2679	0.2222	1.2461	0.2327	0.0030	0.2466
0.1250	0.0000	0.8467	0.0212	1.3699	0.2100	0.0001	0.1321
0.7500	0.7438	0.1181	0.1049	1.2083	0.2578	0.0282	0.3174
0.2545	0.7263	0.2787	0.2222	1.1378	0.2586	0.0349	0.3238
0.8938	0.4625	0.3559	0.0001	1.2152	0.1522	0.0140	0.2960
1.0000	0.6250	0.0831	0.0003	1.2176	0.0955	0.0015	0.2242
0.7844	0.5250	0.3500	0.0021	1.1939	0.2215	0.0640	0.3418
0.9063	0.6875	0.0943	0.0042	1.2145	0.1497	0.0058	0.2680
0.8859	0.5758	0.1981	0.0008	1.2056	0.1577	0.0145	0.2972
0.5375	0.7500	0.0842	0.2175	1.1755	0.3523	0.0914	0.3522
0.1886	0.4341	0.6915	0.2196	1.1520	0.2049	0.0428	0.3299
0.3112	0.4806	0.4971	0.2122	1.1082	0.2784	0.1607	0.3684
0.4943	0.3652	0.4735	0.0198	1.1972	0.3394	0.2306	0.3786
0.3528	0.5875	0.3139	0.2190	1.0840	0.3010	0.1841	0.3722
0.3750	1.0000	0.1842	0.2222	1.3542	0.3955	0.0015	0.2242
0.3125	0.9333	0.2109	0.2222	1.3034	0.3366	0.0037	0.2531
0.8441	0.8987	0.0676	0.1617	1.2570	0.2146	0.0012	0.2161
0.3954	0.4113	0.5277	0.1269	1.1525	0.3179	0.2278	0.3782
0.3899	0.3149	0.7178	0.0457	1.2291	0.3244	0.1303	0.3624
1.0000	0.5000	0.1610	0.0000	1.2199	0.0940	0.0021	0.2349
1.0000	1.0000	0.0359	0.1111	1.2712	0.1207	0.0000	0.0386
0.2500	0.0000	0.8189	0.0024	1.4069	0.3119	0.0006	0.1911
0.7500	0.0000	0.2521	0.0000	1.3150	0.3119	0.0006	0.1911
1.0000	0.2500	0.2302	0.0000	1.2573	0.1001	0.0006	0.1911
1.0000	0.7500	0.0504	0.0024	1.2295	0.1001	0.0006	0.1911
0.7500	1.0000	0.0810	0.2198	1.2913	0.3119	0.0006	0.1911
0.2500	1.0000	0.2222	0.2222	1.3496	0.3119	0.0006	0.1911
0.0000	0.7500	0.3395	0.2222	1.1987	0.1001	0.0006	0.1911
0.0000	0.2500	0.8026	0.2198	1.2683	0.1001	0.0006	0.1911
0.8750	1.0000	0.0553	0.2010	1.2791	0.2100	0.0001	0.1321
1.0000	0.3750	0.2405	0.0000	1.2354	0.0955	0.0015	0.2242
1.0000	0.8750	0.0406	0.0212	1.2504	0.1082	0.0001	0.1321
0.6250	1.0000	0.1124	0.2220	1.3099	0.3955	0.0015	0.2242
0.0000	0.3750	0.6464	0.2220	1.2134	0.0955	0.0015	0.2242
0.0000	0.1250	0.8220	0.2010	1.3151	0.1082	0.0001	0.1321
0.5625	1.0000	0.1292	0.2221	1.3218	0.4196	0.0019	0.2323
0.0000	0.4375	0.5564	0.2221	1.1907	0.0944	0.0019	0.2323
0.6500	0.4875	0.3944	0.0113	1.1735	0.2975	0.2107	0.3761

Table 4. Cont.

X	Y	$F_1(x,y)$	$F_2(x,y)$	$F_3(x,y)$	$F_4(x,y)$	$F_5(x,y)$	$F_6(x,y)$
0.0811	0.5625	0.4329	0.2222	1.1446	0.1376	0.0088	0.2815
0.1154	0.6538	0.3598	0.2222	1.1420	0.1614	0.0103	0.2865
0.3750	0.0000	0.6331	0.0003	1.4141	0.3955	0.0015	0.2242
0.3181	0.1035	0.9263	0.0046	1.3910	0.3299	0.0071	0.2745
0.5197	0.1873	0.5032	0.0006	1.3174	0.3669	0.0457	0.3318
0.4371	0.1016	0.6122	0.0005	1.3796	0.3829	0.0124	0.2921
0.1625	0.2125	1.1918	0.1580	1.3042	0.2034	0.0062	0.2704
0.2375	0.2875	1.1146	0.1580	1.2528	0.2460	0.0331	0.3222
0.0625	0.3125	0.8667	0.2198	1.2383	0.1310	0.0034	0.2507
0.7625	0.2625	0.6033	0.0000	1.2596	0.2488	0.0264	0.3154
0.6298	0.3677	0.5269	0.0020	1.2125	0.3115	0.1663	0.3694
0.5495	0.5455	0.2811	0.1071	1.1349	0.3299	0.3042	0.3863
0.4999	0.6340	0.1996	0.2040	1.1219	0.3394	0.2317	0.3787
0.5581	0.4443	0.3918	0.0254	1.1656	0.3287	0.2924	0.3852
0.4125	0.7796	0.0310	0.2219	1.1740	0.3467	0.0586	0.3392

5. Numerical Results

Delaunay triangulation for all datasets are shown in Figure 8. Meanwhile, Figures 9–11 show the 3D linear interpolant for the scattered datasets, respectively. The surface interpolation by using cubic Timmer triangular patches with selected schemes are shown in Figures 12–23.

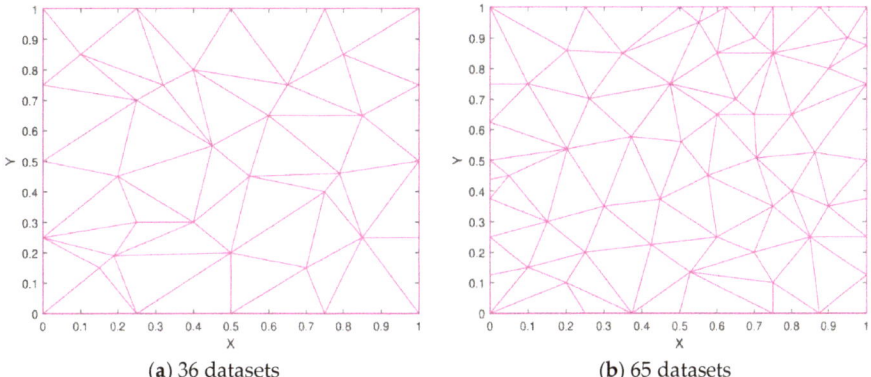

(**a**) 36 datasets (**b**) 65 datasets

Figure 8. Cont.

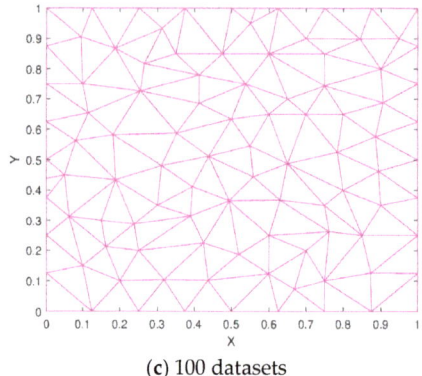

(c) 100 datasets

Figure 8. Delaunay triangulation.

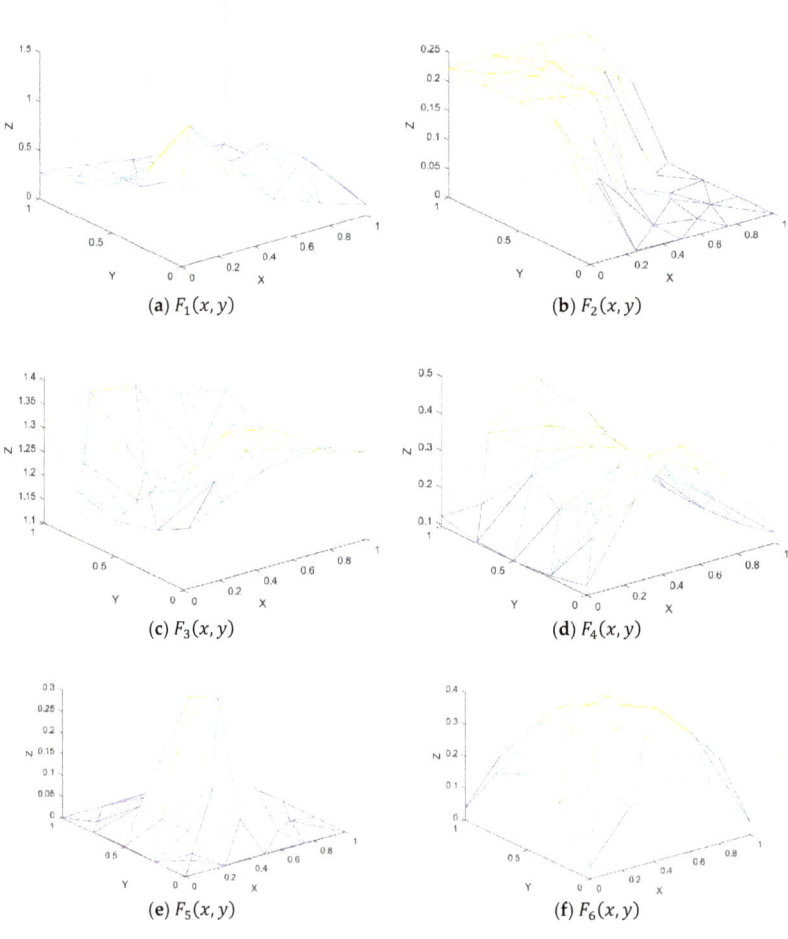

(a) $F_1(x,y)$

(b) $F_2(x,y)$

(c) $F_3(x,y)$

(d) $F_4(x,y)$

(e) $F_5(x,y)$

(f) $F_6(x,y)$

Figure 9. 3D linear interpolant for 36 datasets.

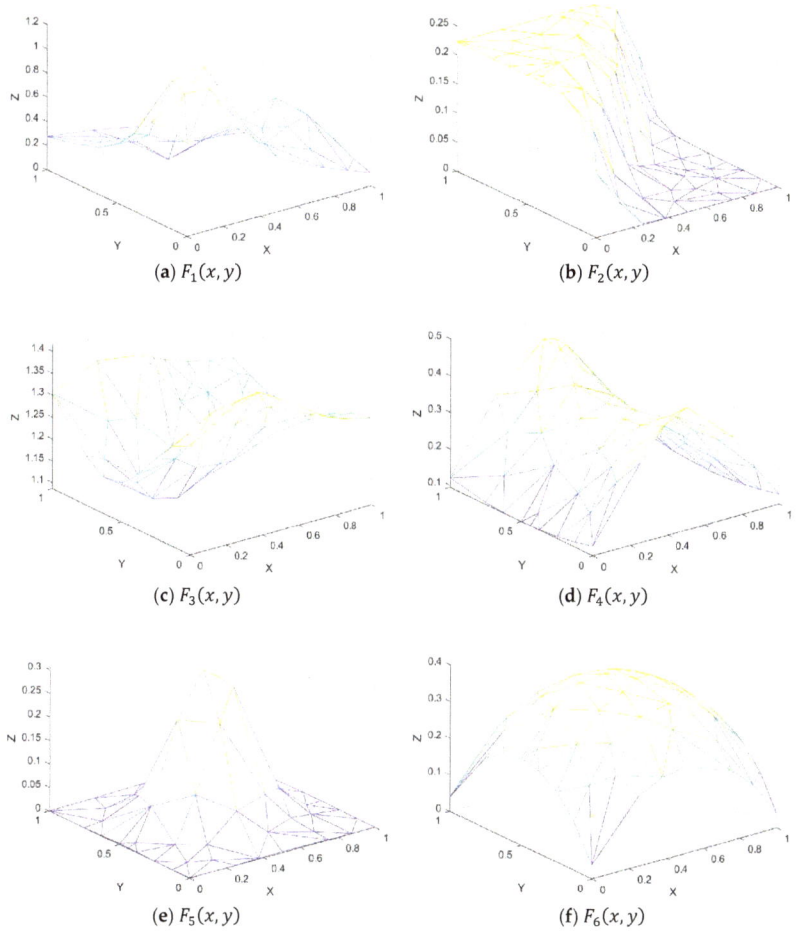

Figure 10. 3D linear interpolant for 65 datasets.

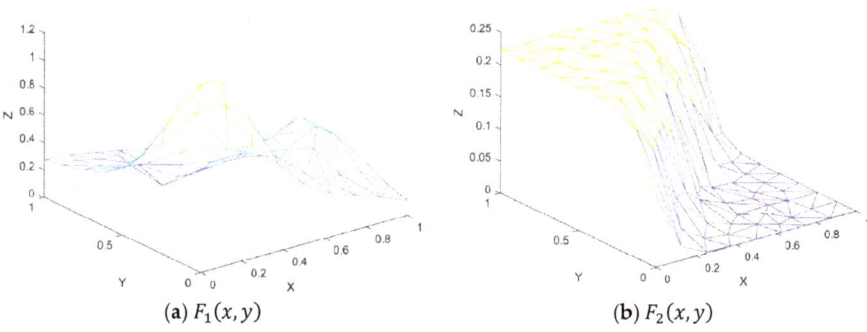

Figure 11. *Cont.*

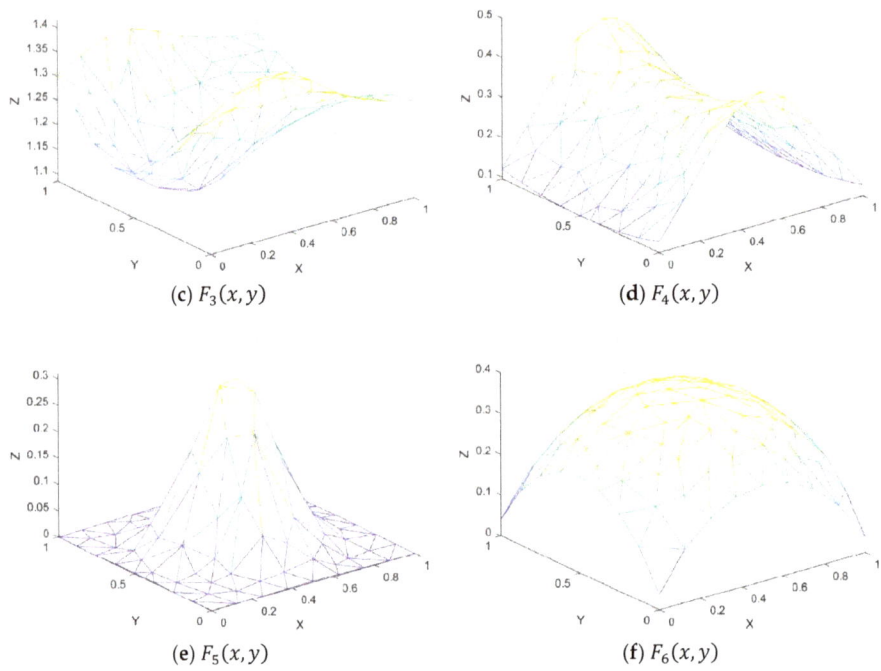

(c) $F_3(x,y)$ (d) $F_4(x,y)$

(e) $F_5(x,y)$ (f) $F_6(x,y)$

Figure 11. 3D linear interpolant for 100 datasets.

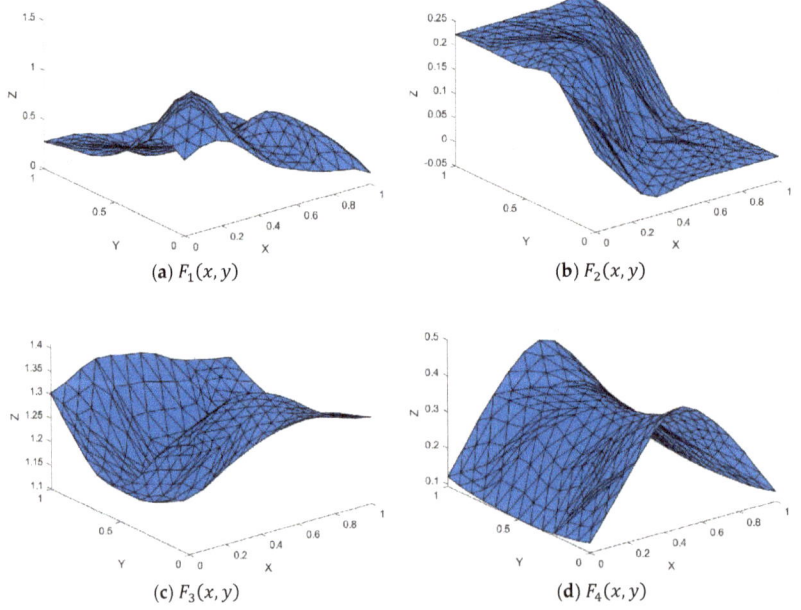

(a) $F_1(x,y)$ (b) $F_2(x,y)$

(c) $F_3(x,y)$ (d) $F_4(x,y)$

Figure 12. *Cont.*

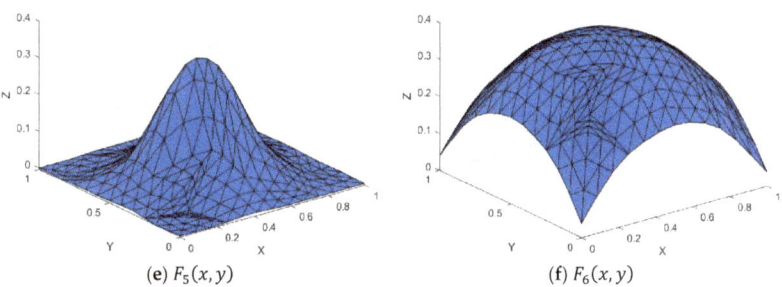

(e) $F_5(x, y)$ (f) $F_6(x, y)$

Figure 12. Surface interpolation using Goodman and Said method and Choice 1 for 36 datasets.

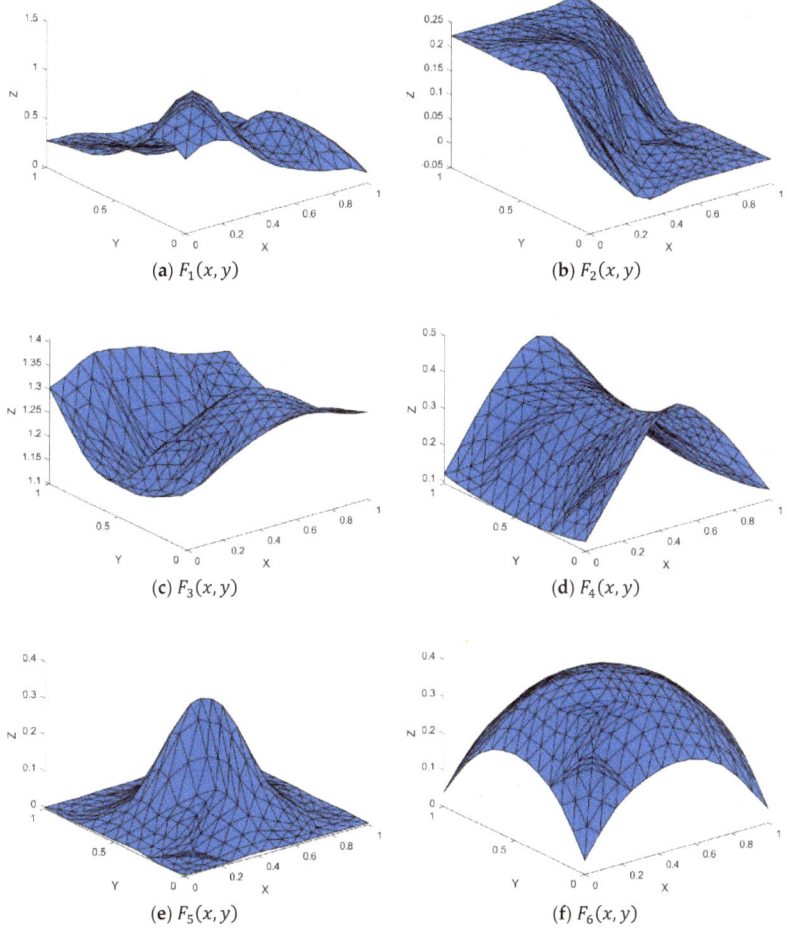

(a) $F_1(x, y)$ (b) $F_2(x, y)$

(c) $F_3(x, y)$ (d) $F_4(x, y)$

(e) $F_5(x, y)$ (f) $F_6(x, y)$

Figure 13. Surface interpolation using Goodman and Said method and Choice 2 for 36 datasets.

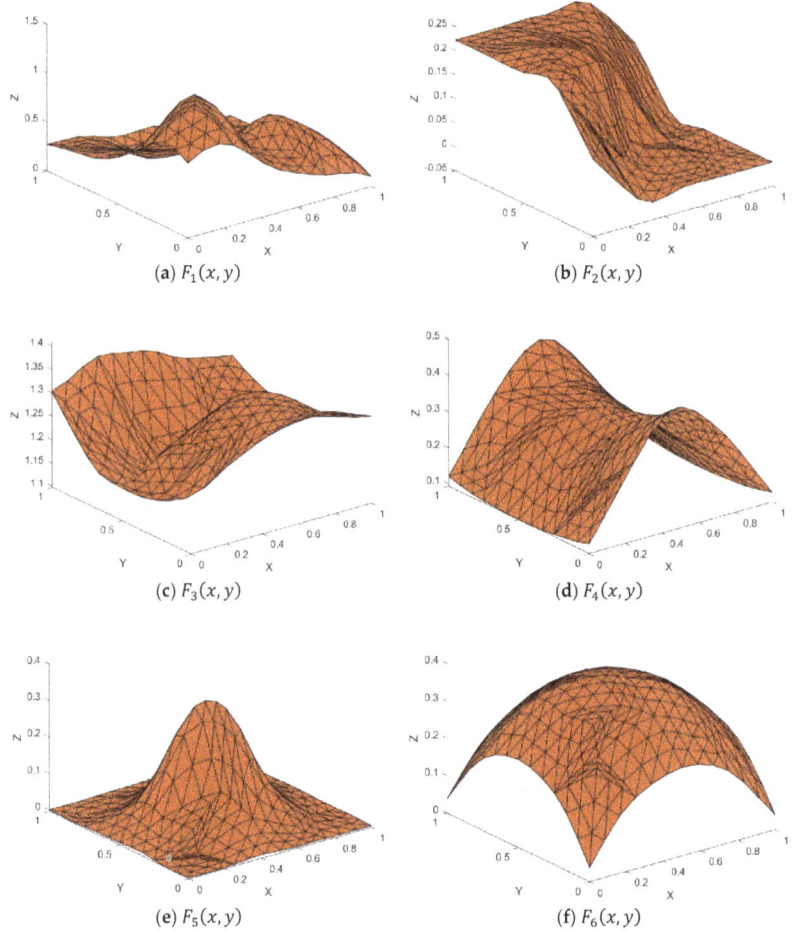

Figure 14. Surface interpolation using Foley and Opitz method and Choice 1 for 36 datasets.

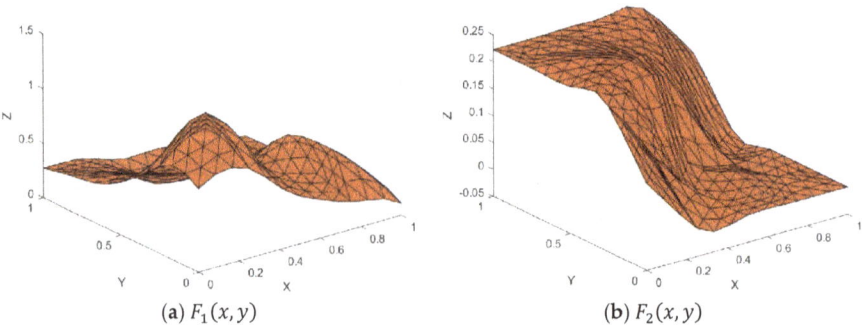

Figure 15. *Cont.*

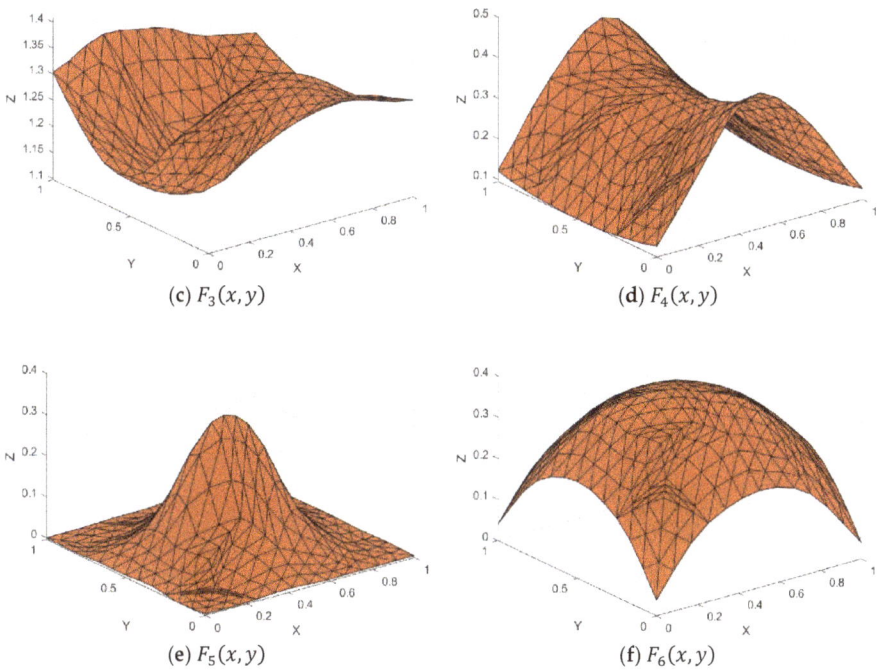

Figure 15. Surface interpolation using Foley and Opitz method and Choice 2 for 36 datasets.

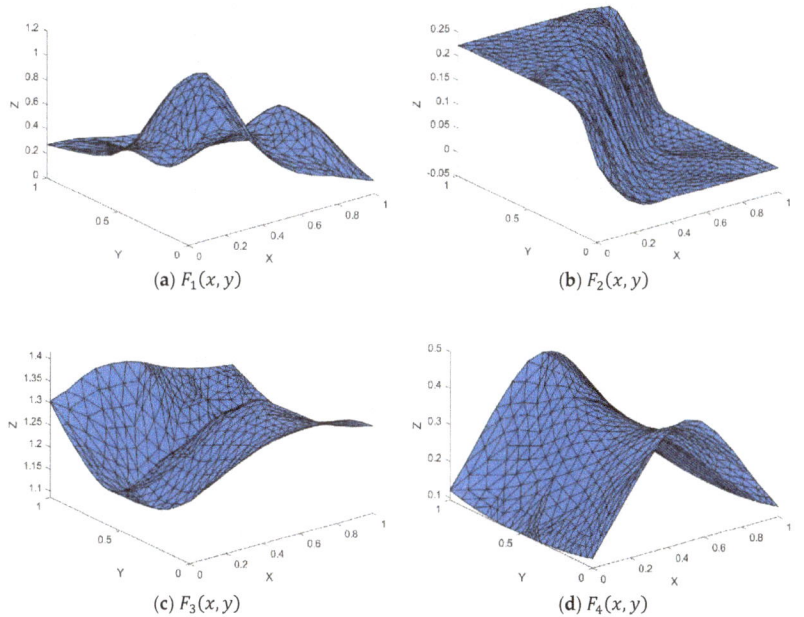

Figure 16. *Cont.*

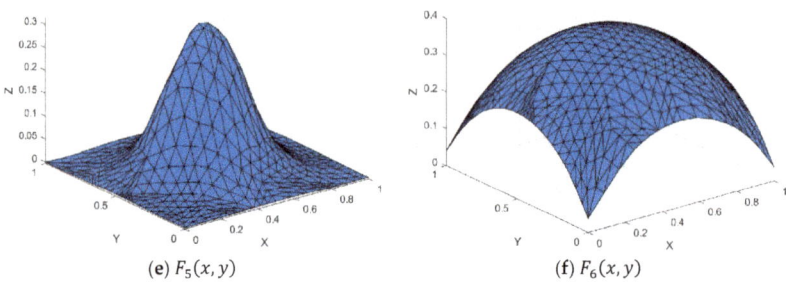

(e) $F_5(x, y)$ (f) $F_6(x, y)$

Figure 16. Surface interpolation using Goodman and Said method and Choice 1 for 65 datasets.

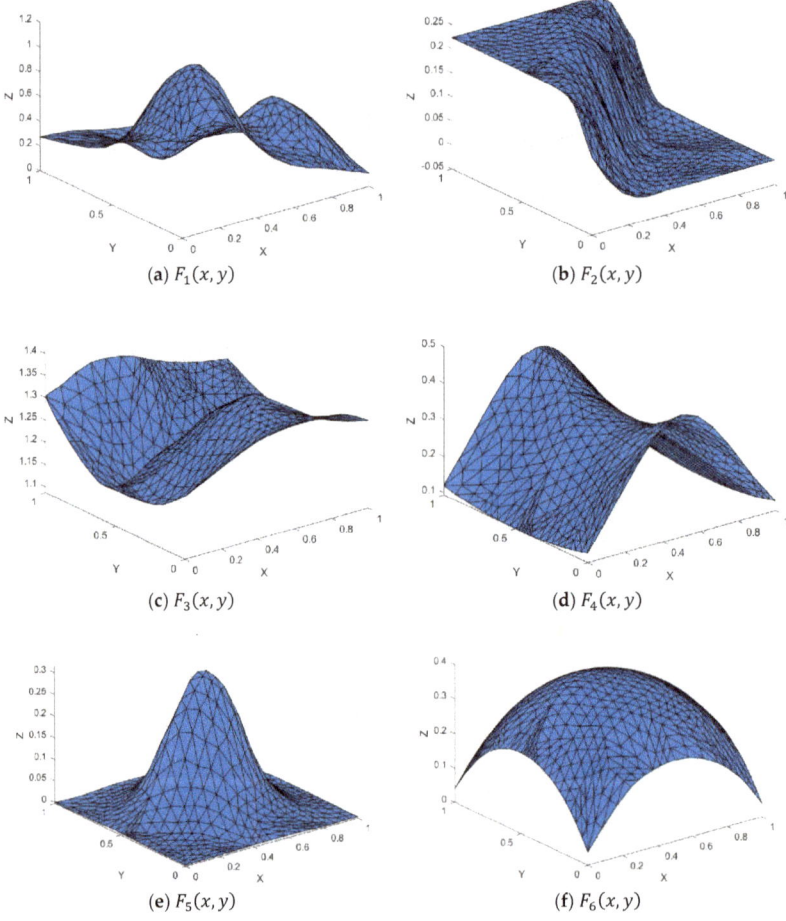

(a) $F_1(x, y)$ (b) $F_2(x, y)$

(c) $F_3(x, y)$ (d) $F_4(x, y)$

(e) $F_5(x, y)$ (f) $F_6(x, y)$

Figure 17. Surface interpolation using Goodman and Said method and Choice 2 for 65 datasets.

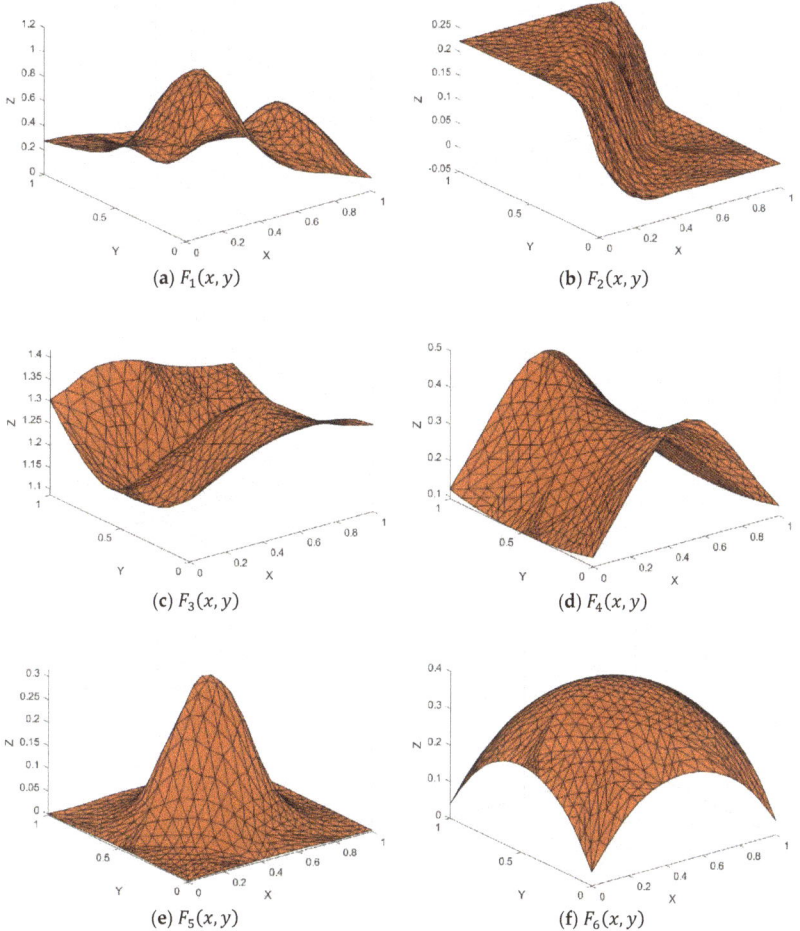

Figure 18. Surface interpolation using Foley and Opitz method and Choice 1 for 65 datasets.

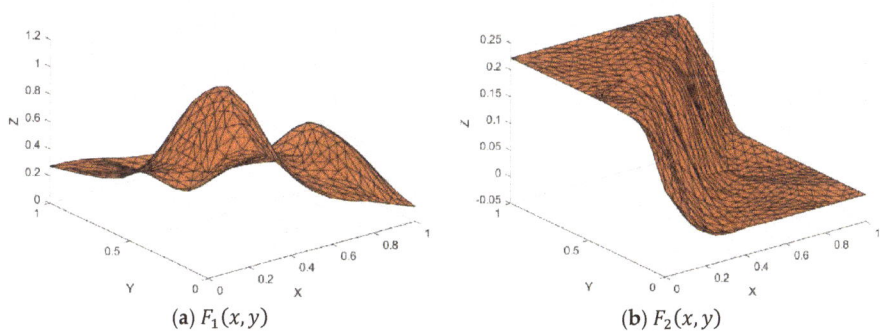

Figure 19. Cont.

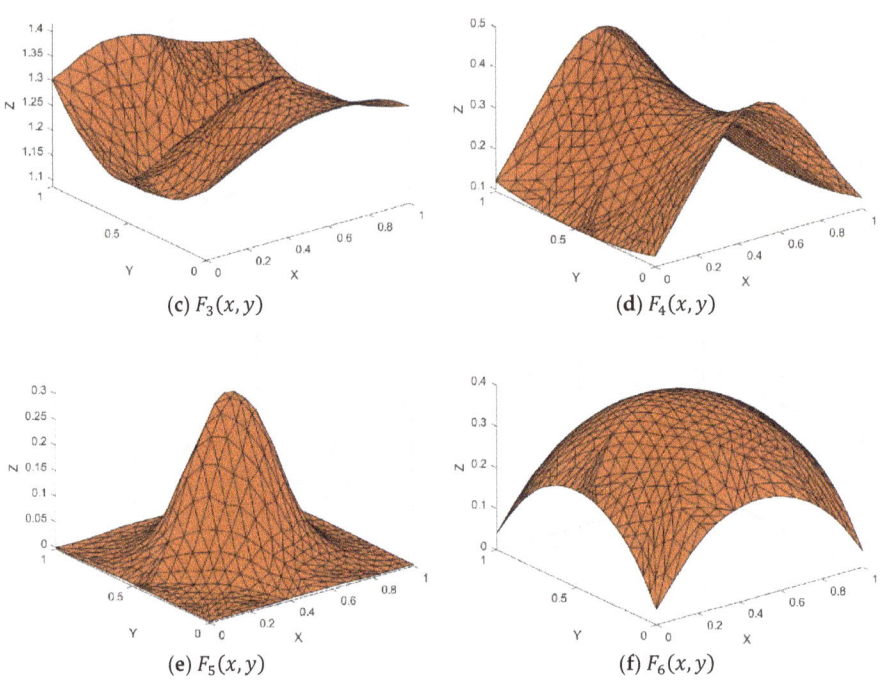

Figure 19. Surface interpolation using Foley and Opitz method and Choice 2 for 65 datasets.

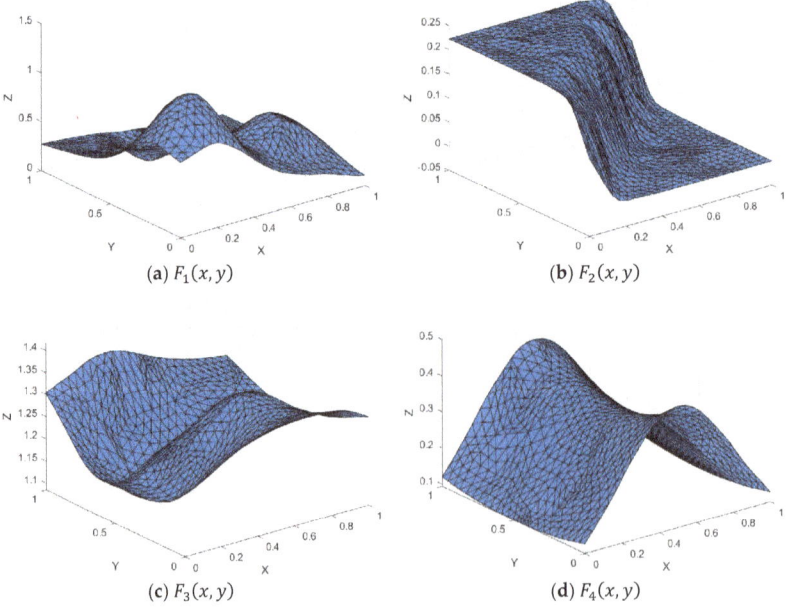

Figure 20. *Cont.*

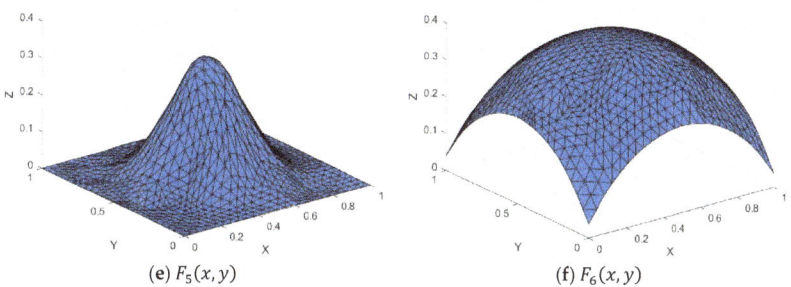

(e) $F_5(x,y)$ (f) $F_6(x,y)$

Figure 20. Surface interpolation using Goodman and Said method and Choice 1 for 100 datasets.

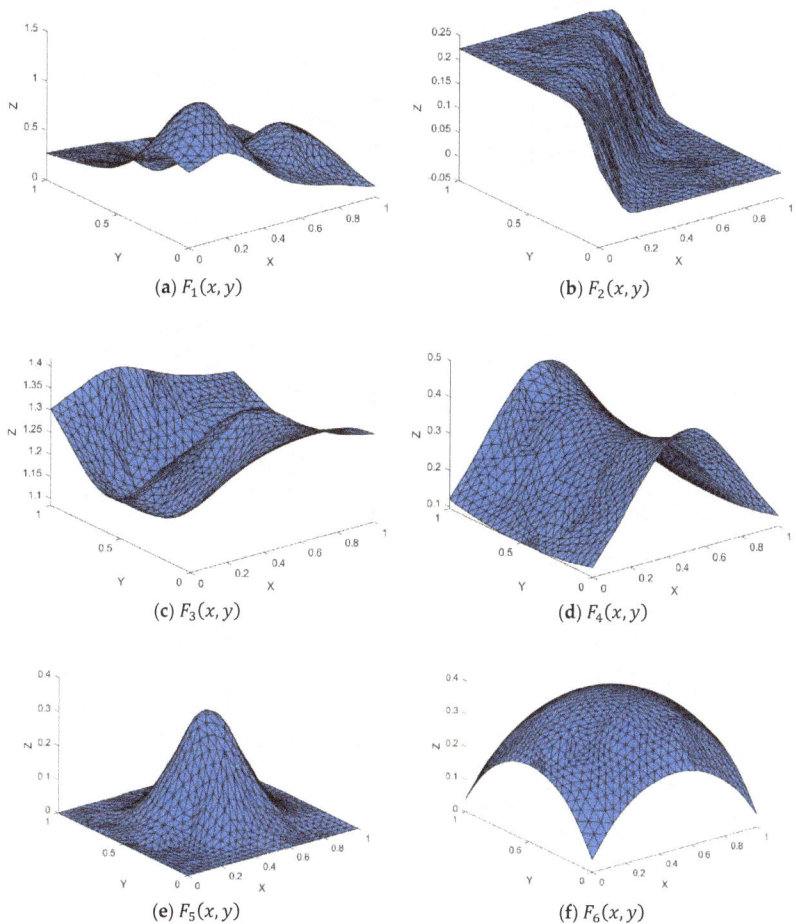

(a) $F_1(x,y)$ (b) $F_2(x,y)$

(c) $F_3(x,y)$ (d) $F_4(x,y)$

(e) $F_5(x,y)$ (f) $F_6(x,y)$

Figure 21. Surface interpolation using Goodman and Said method and Choice 2 for 100 datasets.

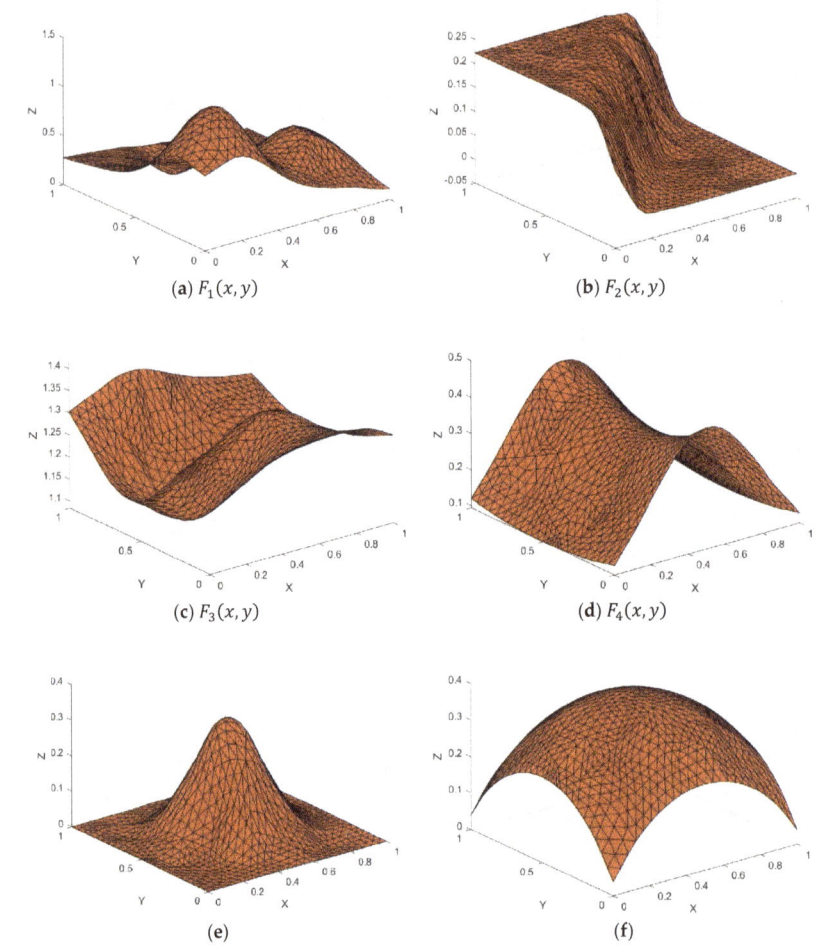

Figure 22. Surface interpolation using Foley and Opitz method and Choice 1 for 100 datasets.

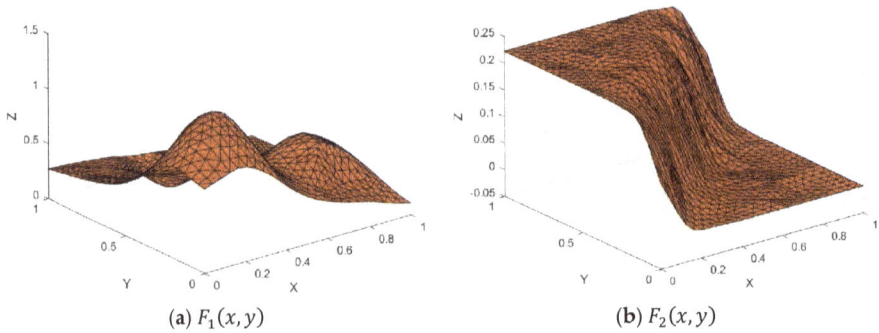

Figure 23. *Cont.*

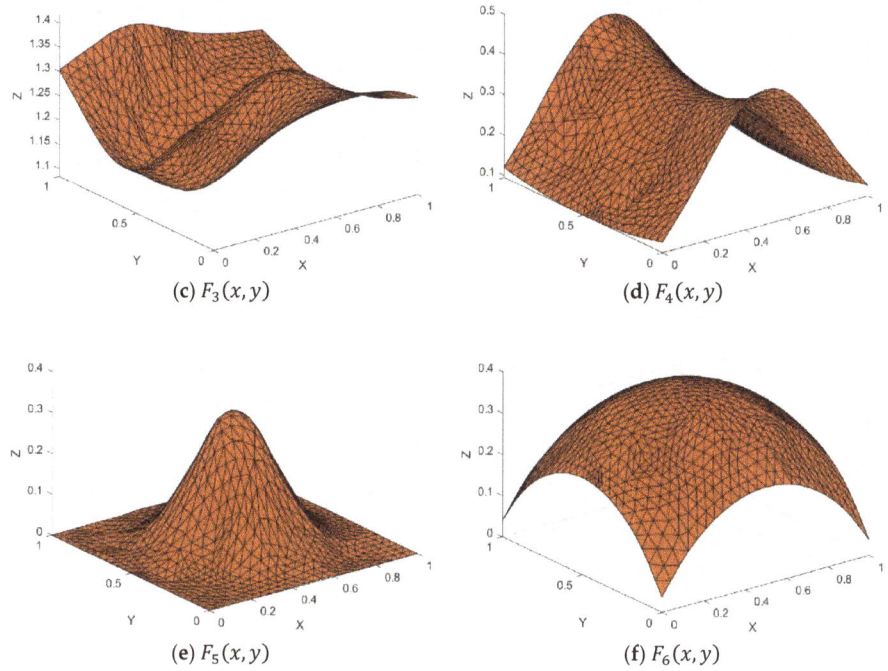

Figure 23. Surface interpolation using Foley and Opitz method and Choice 2 for 100 datasets.

Based on Figures 12–23, all schemes are capable of producing a smooth C^1 surface. The 36, 65, and 100 datasets consist of 54, 100, and 164 triangular patches with C^1 continuity for each edge, respectively. Visually, the proposed scheme produces smooth surfaces for all datasets. However, in order to measure the effectiveness of the proposed scattered data interpolation scheme, we calculate root mean square error (RMSE), maximum error (Max error), coefficient of determination (R^2), and central processor unit (CPU) time in seconds. For the computation time, a comparison has been made between two different methods to calculate the inner ordinates, i.e., Goodman and Said [8] and Foley and Opitz [19] methods and two distinct calculation of local scheme denoted as Choice 1 and Choice 2. The error analysis for 36, 65, and 100 data points are shown in Tables 5–7, respectively.

Table 5. Error analysis for 36 datasets.

Test Function	Error	Goodman and Said		Foley and Opitz	
		Choice 1	Choice 2	Choice 1	Choice 2
$F_1(x,y)$	RMSE	0.025827	0.025901	0.026097	0.026458
	R^2	0.991858	0.991811	0.991687	0.991455
	Max error	0.109091	0.110443	0.109642	0.110142
	CPU time	3.666885	3.792837	3.531956	3.612421
$F_2(x,y)$	RMSE	0.013091	0.013105	0.012899	0.012970
	R^2	0.982658	0.982620	0.983163	0.982978
	Max error	0.049755	0.050222	0.048333	0.049466
	CPU time	3.465182	3.581924	3.238475	3.418465

Table 5. Cont.

Test Function	Error	Goodman and Said		Foley and Opitz	
		Choice 1	Choice 2	Choice 1	Choice 2
$F_3(x,y)$	RMSE	0.006086	0.006097	0.005834	0.005897
	R^2	0.993518	0.993493	0.994043	0.993914
	Max error	0.026836	0.026923	0.026082	0.026475
	CPU time	3.381397	3.412042	3.362575	3.381239
$F_4(x,y)$	RMSE	0.127280	0.127285	0.127392	0.127379
	R^2	0.999012	0.998742	0.998671	0.998923
	Max error	0.335392	0.335392	0.335392	0.335392
	CPU time	3.718293	3.819244	3.634124	3.801234
$F_5(x,y)$	RMSE	0.006244	0.006326	0.005952	0.006095
	R^2	0.993218	0.993039	0.993837	0.993539
	Max error	0.030589	0.031475	0.027931	0.028902
	CPU time	3.931621	4.012938	3.544473	3.712645
$F_6(x,y)$	RMSE	0.003587	0.003602	0.003980	0.003995
	R^2	0.997647	0.997628	0.997104	0.997082
	Max error	0.015551	0.015589	0.015264	0.014997
	CPU time	3.910817	3.912341	3.722930	3.800012

Table 6. Error analysis for 65 datasets.

Test Function	Error	Goodman and Said		Foley and Opitz	
		Choice 1	Choice 2	Choice 1	Choice 2
$F_1(x,y)$	RMSE	0.015504	0.015586	0.015090	0.015246
	R^2	0.997066	0.997034	0.997221	0.997163
	Max error	0.062431	0.063216	0.063829	0.066465
	CPU time	9.012512	9.214511	8.912451	9.109472
$F_2(x,y)$	RMSE	0.005162	0.005177	0.005015	0.005016
	R^2	0.997304	0.997288	0.997457	0.997454
	Max error	0.030704	0.030792	0.031553	0.031558
	CPU time	8.729935	9.497846	8.623959	8.761479
$F_3(x,y)$	RMSE	0.003151	0.003163	0.003018	0.003091
	R^2	0.998263	0.998249	0.998406	0.998328
	Max error	0.015383	0.015461	0.015825	0.016667
	CPU time	8.910241	8.948517	8.891025	8.901256
$F_4(x,y)$	RMSE	0.127247	0.127249	0.127274	0.127264
	R^2	0.998901	0.998703	0.998767	0.998513
	Max error	0.333987	0.333987	0.333987	0.333987
	CPU time	9.193316	9.201537	9.001285	9.182451
$F_5(x,y)$	RMSE	0.004951	0.004980	0.004800	0.004877
	R^2	0.995737	0.995686	0.995992	0.995863
	Max error	0.033110	0.033635	0.031342	0.031923
	CPU time	8.901256	8.924514	8.802561	8.856126
$F_6(x,y)$	RMSE	0.001953	0.001959	0.002122	0.002146
	R^2	0.999303	0.999298	0.999177	0.999158
	Max error	0.010602	0.010715	0.009957	0.009725
	CPU time	8.981256	8.992456	8.900128	8.941588

Table 7. Error analysis for 100 datasets.

Test Function	Error	Goodman and Said		Foley and Opitz	
		Choice 1	Choice 2	Choice 1	Choice 2
$F_1(x,y)$	RMSE	0.006834	0.006887	0.006660	0.006804
	R^2	0.999430	0.999421	0.999459	0.999435
	Max error	0.034165	0.034446	0.032402	0.032503
	CPU time	19.245752	19.157941	18.989466	19.015484
$F_2(x,y)$	RMSE	0.003638	0.003644	0.003525	0.003568
	R^2	0.998661	0.998656	0.998742	0.998712
	Max error	0.023940	0.024076	0.024344	0.025017
	CPU time	17.772122	17.387878	17.726120	17.609815
$F_3(x,y)$	RMSE	0.000997	0.001004	0.000960	0.001022
	R^2	0.999826	0.999824	0.999839	0.999817
	Max error	0.004870	0.005009	0.004337	0.004622
	CPU time	17.377007	17.801830	16.248355	16.449573
$F_4(x,y)$	RMSE	0.127213	0.127209	0.127199	0.127200
	R^2	0.997790	0.997735	0.997851	0.997732
	Max error	0.334366	0.334387	0.333155	0.333431
	CPU time	18.517801	17.531418	17.221382	17.582176
$F_5(x,y)$	RMSE	0.002273	0.002299	0.002159	0.002221
	R^2	0.999101	0.999081	0.999189	0.999142
	Max error	0.012672	0.012685	0.012559	0.012585
	CPU time	19.092384	19.181738	18.463083	19.083249
$F_6(x,y)$	RMSE	0.001116	0.001153	0.000867	0.000868
	R^2	0.999772	0.999757	0.999863	0.999862
	Max error	0.007914	0.007917	0.005458	0.005459
	CPU time	18.521953	19.728907	17.658403	18.307126

Based on Tables 5–7, the numerical results obtained by using the local scheme with Choice 2 gave larger error than Choice 1 while most of the Foley and Opitz [19] method gave smaller error than the Goodman and Said [8] method. In terms of CPU time (in seconds), the Goodman and Said [8] method takes more time than the Foley and Opitz [19] method. Convex scheme using Choice 2 took longer time than scheme with Choice 1. The main reason is because convex combination using Choice 2 requires more calculation than Choice 1. Furthermore, the reason the Foley and Opitz [19] method has less time is that their scheme considers two triangular patches to calculate the inner ordinates while the Goodman and Said [8] method needs to find the three inner ordinates for each triangular patch.

Hence, we conclude that the best scheme for scattered data interpolation is the cubic Timmer triangular patches with convex combination of Choice 1 and the Foley and Opitz [19] method to calculate the inner ordinates. Figure 24 shows the error comparison of the proposed cubic Timmer triangular patch with all test functions using the best scheme mentioned above with different datasets.

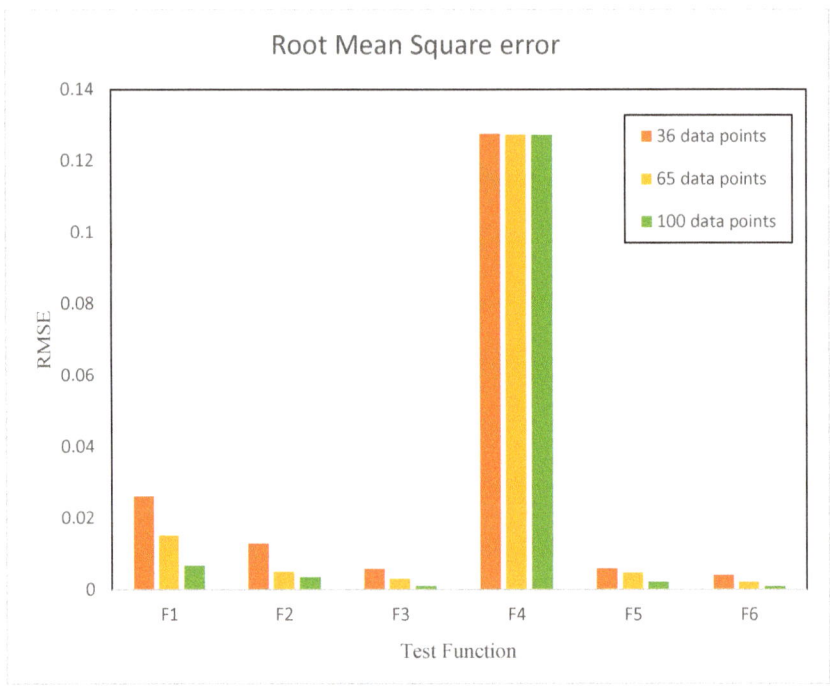

(a) RMSE

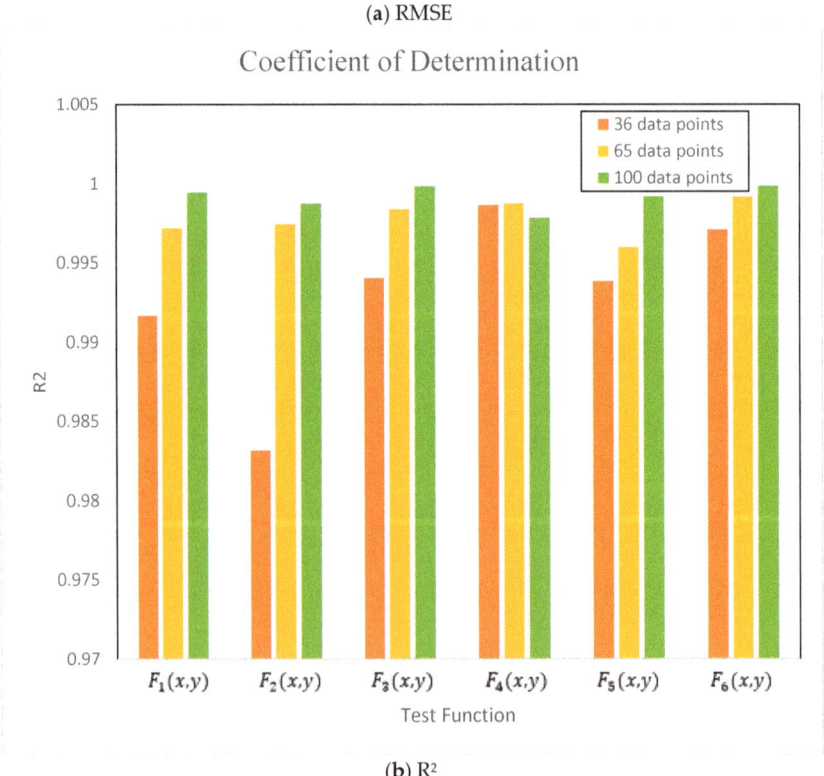

(b) R^2

Figure 24. *Cont.*

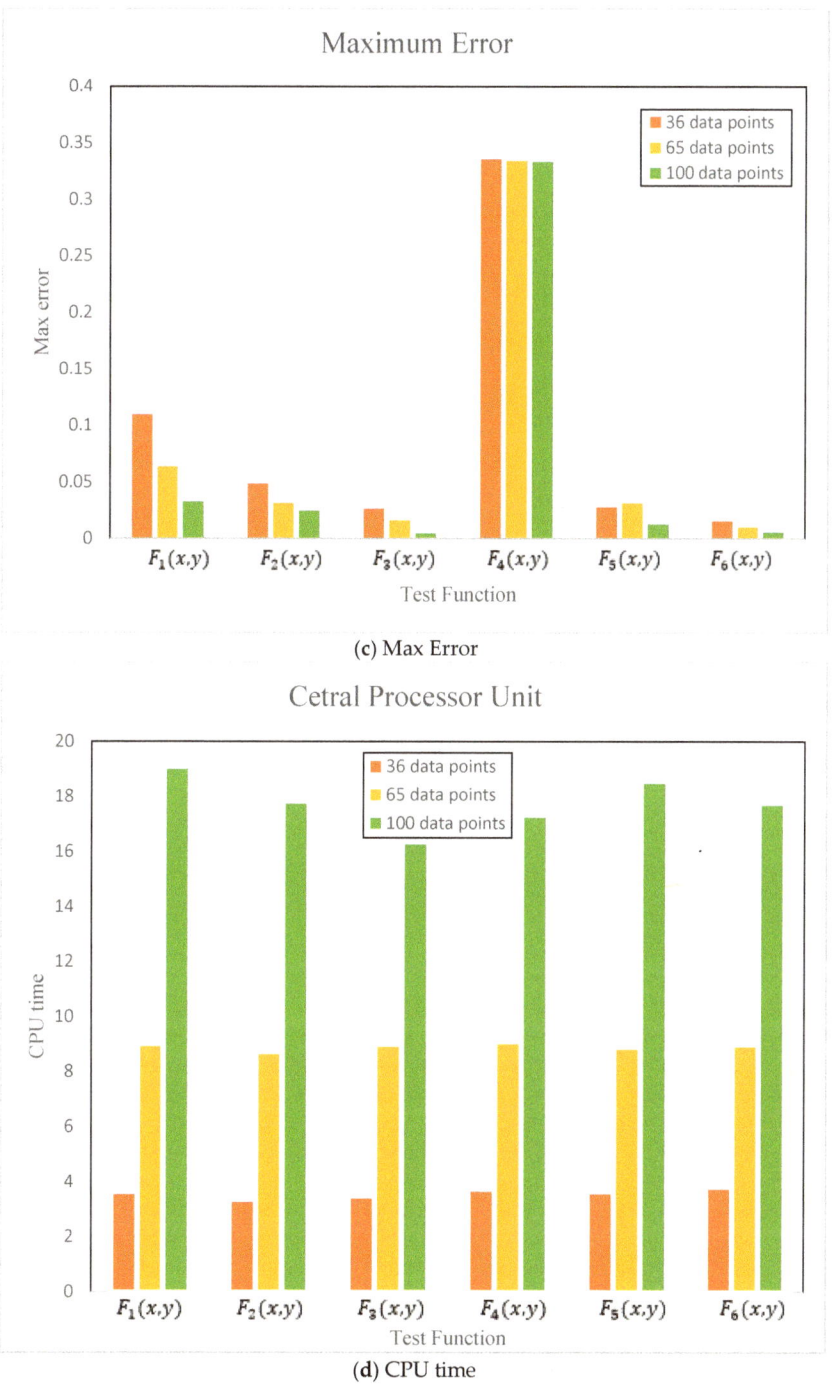

(c) Max Error

(d) CPU time

Figure 24. Comparison of the proposed method using the best schemes.

Based on Figure 24, as the number of data points increased, the errors such as RMSE and Max error will be decreased. For CPU time, when more data are used, it will take a longer time, while the comparison using R^2 shows that when the datasets used increases, the R^2 value will increase too.

Next, we compare the proposed cubic Timmer triangular scheme with the established schemes such as Karim and Saaban [21] and Goodman and Said [8]. We use 100 data points as shown in Table 4. The numerical comparisons are shown in Table 8.

Table 8. Comparison with established schemes.

Test Function	Error	100 Data Points		
		Goodman and Said [8]	Karim and Saaban [21]	Proposed Scheme
$F_1(x,y)$	RMSE	0.006523	0.006543	0.006660
	R^2	0.999481	0.999477	0.999459
	Max error	0.032346	0.033434	0.032402
	CPU time	19.853309	19.740250	18.989466
$F_2(x,y)$	RMSE	0.003486	0.003464	0.003525
	R^2	0.998770	0.998786	0.998742
	Max error	0.023774	0.022634	0.024344
	CPU time	18.454829	17.937833	17.726120
$F_3(x,y)$	RMSE	0.000953	0.001151	0.000960
	R^2	0.999841	0.999768	0.999839
	Max error	0.004293	0.005725	0.004337
	CPU time	16.947552	16.851681	16.748355
$F_4(x,y)$	RMSE	0.127205	0.127190	0.127200
	R^2	0.997654	0.998057	0.997735
	Max error	0.334366	0.334366	0.334366
	CPU time	18.045273	17.586388	17.582176
$F_5(x,y)$	RMSE	0.002093	0.002060	0.002159
	R^2	0.999238	0.999262	0.999189
	Max error	0.012460	0.012264	0.012559
	CPU time	18.675766	18.590993	18.463083
$F_6(x,y)$	RMSE	0.000873	0.000881	0.000867
	R^2	0.999861	0.999858	0.999863
	Max error	0.005458	0.005458	0.005458
	CPU time	17.956438	17.777269	17.658403

From Table 8, based on RMSE, max error, and R^2, we can see clearly that cubic Timmer triangular patch is on par with Karim and Saaban [21] and Goodman and Said [8] schemes. However, in terms of CPU time (in seconds), the proposed scheme require smaller CPU time compared to the other established methods. Thus, we believed that the proposed cubic Timmer triangular patch is suitable to interpolate dense or big scattered datasets, since it requires less CPU time than [8] and [21].

To validate this, we test the proposed cubic Timmer triangular patch scheme by using the seamount dataset obtained in MATLAB. The seamount dataset represents the surface of underwater mountain that is located at 48.2°S, 148.8°W on the Louisville Ridge in the South Pacific in 1984. The seamount data et contains 294 data points and it consists a set of longitude (X), latitude (Y), and depth-in-feet (Z), as shown in Table 9. Table 9 shows 294 data points of the seamount dataset. There are about 566 triangles. Figures 25 and 26 show the Delaunay triangulation and the 3D visualization of the seamount dataset, respectively. The surface interpolation of 566 triangular patches formed by using the proposed cubic Timmer, cubic Ball, and cubic Bèzier are shown in Figure 27.

Table 9. Seamount dataset.

X	Y	Z	X	Y	Z	X	Y	Z
211.18	−47.97	−4250	211.07	−48.18	−3050	211.19	−48.215	−1100
211.28	−47.97	−4250	211.11	−48.18	−2615	211.2	−48.215	−1000
211.1	−47.98	−4250	211.15	−48.18	−1850	211.21	−48.215	−975
211.38	−47.98	−4250	211.17	−48.18	−1730	211.22	−48.215	−925
211.45	−48	−4250	211.19	−48.18	−1150	211.23	−48.215	−725
211.23	−48.01	−3900	211.2	−48.18	−1025	211.24	−48.215	−800
211.31	−48.01	−3950	211.21	−48.18	−600	211.25	−48.215	−1050
210.98	−48.02	−4250	211.22	−48.18	−900	210.95	−48.22	−4050
211.13	−48.02	−3900	211.23	−48.18	−1050	211.07	−48.22	−3700
211.39	−48.02	−4000	211.24	−48.18	−800	211.11	−48.22	−2910
211.06	−48.03	−4050	211.25	−48.18	−950	211.15	−48.22	−2150
211.51	−48.03	−4250	211.26	−48.18	−1300	211.17	−48.22	−1750
211.19	−48.04	−3700	211.27	−48.18	−1370	211.19	−48.22	−1250
211.26	−48.04	−3730	211.28	−48.18	−1450	211.2	−48.22	−1150
211.32	−48.05	−3650	211.29	−48.18	−1490	211.21	−48.22	−1125
211.1	−48.06	−3800	211.31	−48.18	−1850	211.22	−48.22	−950
211.01	−48.07	−3980	211.34	−48.18	−2575	211.23	−48.22	−950
211.39	−48.07	−3700	211.38	−48.18	−3350	211.24	−48.22	−925
211.48	−48.07	−3980	211.19	−48.185	−1300	211.25	−48.22	−1125
211.17	−48.08	−3280	211.2	−48.185	−1050	211.27	−48.22	−1350
211.21	−48.08	−3100	211.21	−48.185	−650	211.29	−48.22	−1650
211.25	−48.08	−3140	211.22	−48.185	−770	211.31	−48.22	−1750
211.29	−48.08	−3250	211.23	−48.185	−750	211.34	−48.22	−2500
210.91	−48.09	−4250	211.24	−48.185	−620	211.38	−48.22	−3025
211.57	−48.09	−4250	211.25	−48.185	−950	211.42	−48.22	−3400
211.15	−48.1	−3150	211.26	−48.185	−1150	211.16	−48.23	−2200
211.19	−48.1	−3000	211.27	−48.185	−1000	211.18	−48.23	−1850
211.23	−48.1	−2850	211.28	−48.185	−1150	211.2	−48.23	−1500
211.27	−48.1	−3000	210.89	−48.19	−4250	211.22	−48.23	−1325
211.31	−48.1	−3100	211	−48.19	−3650	211.24	−48.23	−1375
211.34	−48.1	−3220	211.14	−48.19	−2300	211.26	−48.23	−1530
211.06	−48.11	−3630	211.16	−48.19	−1940	211.28	−48.23	−1680
211.47	−48.11	−3765	211.18	−48.19	−1550	211.3	−48.23	−2000
211.13	−48.12	−3170	211.2	−48.19	−1050	211.02	−48.24	−3700
211.17	−48.12	−2875	211.21	−48.19	−675	211.09	−48.24	−3325
211.21	−48.12	−2600	211.22	−48.19	−600	211.13	−48.24	−2875
211.25	−48.12	−2600	211.23	−48.19	−590	211.17	−48.24	−2200
211.29	−48.12	−2575	211.24	−48.19	−650	211.19	−48.24	−1850
211.32	−48.12	−2950	211.25	−48.19	−800	211.21	−48.24	−1600
211.53	−48.12	−4070	211.26	−48.19	−1050	211.23	−48.24	−1900
210.96	−48.14	−3920	211.27	−48.19	−950	211.25	−48.24	−1800
211.11	−48.14	−2950	211.28	−48.19	−1000	211.27	−48.24	−1930
211.15	−48.14	−2550	211.3	−48.19	−1780	211.29	−48.24	−2000
211.19	−48.14	−2350	211.19	−48.19	−1150	211.31	−48.24	−2250
211.23	−48.14	−2195	211.2	−48.195	−850	211.36	−48.24	−2800
211.27	−48.14	−2080	211.21	−48.195	−600	211.4	−48.24	−3220
211.3	−48.14	−2450	211.22	−48.195	−570	211.44	−48.24	−3500
211.34	−48.14	−2925	211.23	−48.195	−555	211.53	−48.24	−3650
211.38	−48.14	−3125	211.24	−48.195	−580	211.65	−48.24	−4250
211.04	−48.15	−3450	211.25	−48.195	−700	211.18	−48.25	−2150
211.16	−48.15	−2110	211.26	−48.195	−750	211.2	−48.25	−1840
211.18	−48.15	−2100	211.27	−48.195	−875	211.22	−48.25	−2275
211.2	−48.15	−1760	211.28	−48.195	−1020	211.24	−48.25	−2275
211.22	−48.15	−1920	211.05	−48.2	−3275	211.26	−48.25	−2150
211.24	−48.15	−1900	211.09	−48.2	−2865	211.28	−48.25	−2250
211.26	−48.15	−1750	211.13	−48.2	−2480	210.93	−48.26	−4250
211.28	−48.15	−2110	211.15	−48.2	−2025	211.11	−48.26	−3240
211.51	−48.15	−3950	211.17	−48.2	−1375	211.15	−48.26	−2675

137

Table 9. *Cont.*

X	Y	Z	X	Y	Z	X	Y	Z
211.6	−48.15	−4250	211.18	−48.2	−1000	211.19	−48.26	−2100
211.09	−48.16	−2950	211.19	−48.2	−825	211.23	−48.26	−2575
211.13	−48.16	−2570	211.2	−48.2	−700	211.27	−48.26	−2400
211.15	−48.16	−1950	211.21	−48.2	−580	211.3	−48.26	−2550
211.17	−48.16	−1750	211.22	−48.2	−510	211.34	−48.26	−2820
211.19	−48.16	−1480	211.23	−48.2	−500	211.38	−48.26	−3050
211.2	−48.16	−1325	211.24	−48.2	−550	211.42	−48.26	−3400
211.21	−48.16	−1350	211.25	−48.2	−600	211.06	−48.28	−3725
211.23	−48.16	−1650	211.26	−48.2	−735	211.13	−48.28	−3120
211.25	−48.16	−1375	211.27	−48.2	−875	211.17	−48.28	−2800
211.27	−48.16	−1780	211.28	−48.2	−1150	211.21	−48.28	−3050
211.29	−48.16	−2125	211.29	−48.2	−1500	211.25	−48.28	−2925
211.31	−48.16	−2200	211.31	−48.2	−2150	211.28	−48.28	−2775
211.36	−48.16	−2940	211.36	−48.2	−3000	211.32	−48.28	−2920
211.42	−48.16	−3450	211.4	−48.2	−3380	211.36	−48.28	−3190
211.19	−48.165	−1150	211.48	−48.2	−3780	211.4	−48.28	−3260
211.2	−48.165	−1125	211.57	−48.2	−4025	211.49	−48.28	−3780
211.21	−48.165	−1150	211.18	−48.205	−950	211.59	−48.28	−4050
211.25	−48.165	−1125	211.19	−48.205	−800	211.26	−48.3	−3250
211.14	−48.17	−2250	211.2	−48.205	−740	211.3	−48.3	−3140
211.16	−48.17	−1875	211.21	−48.205	−595	210.99	−48.32	−4250
211.18	−48.17	−1340	211.22	−48.205	−595	211.08	−48.32	−3950
211.19	−48.17	−1075	211.23	−48.205	−490	211.15	−48.32	−3750
211.2	−48.17	−850	211.24	−48.205	−650	211.22	−48.32	−3630
211.21	−48.17	−850	211.25	−48.205	−748	211.28	−48.32	−3420
211.22	−48.17	−1100	211.26	−48.205	−850	211.36	−48.32	−3420
211.23	−48.17	−1375	211.27	−48.205	−1000	211.46	−48.32	−3735
211.24	−48.17	−1175	211.16	−48.21	−1700	211.56	−48.32	−4015
211.25	−48.17	−950	211.18	−48.21	−1200	211.66	−48.32	−4250
211.26	−48.17	−1300	211.19	−48.21	−1000	211.18	−48.35	−4010
211.28	−48.17	−1825	211.2	−48.21	−850	211.1	−48.37	−4250
211.3	−48.17	−1850	211.21	−48.21	−765	211.26	−48.37	−3950
211.32	−48.17	−2110	211.22	−48.21	−780	211.34	−48.37	−3850
211.19	−48.175	−1125	211.23	−48.21	−560	211.42	−48.37	−3900
211.2	−48.175	−800	211.24	−48.21	−750	211.5	−48.37	−4050
211.21	−48.175	−700	211.25	−48.21	−850	211.6	−48.39	−4250
211.22	−48.175	−1020	211.26	−48.21	−1040	211.22	−48.4	−4250
211.23	−48.175	−1175	211.27	−48.21	−1200	211.3	−48.42	−4250
211.24	−48.175	−900	211.28	−48.21	−1300	211.38	−48.42	−4250
211.25	−48.175	−900	211.3	−48.21	−1600			

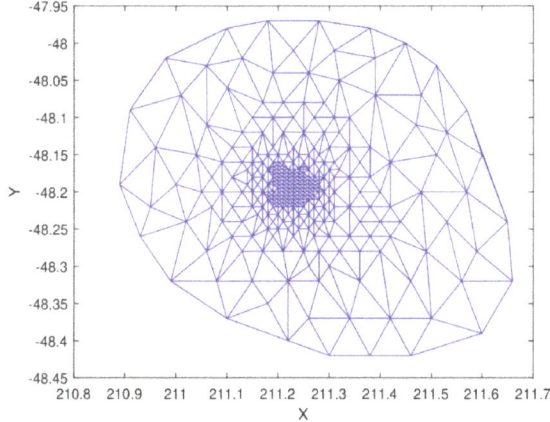

Figure 25. Delaunay triangulation of seamount dataset.

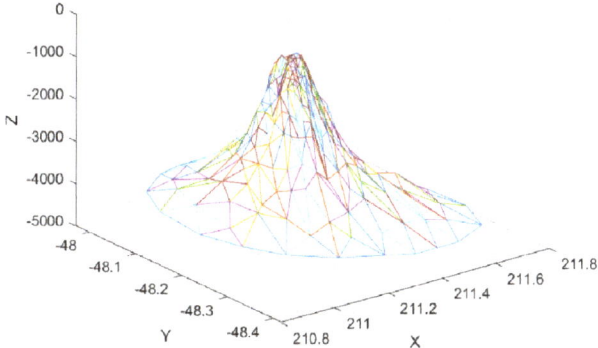

Figure 26. 3D visualization of seamount dataset.

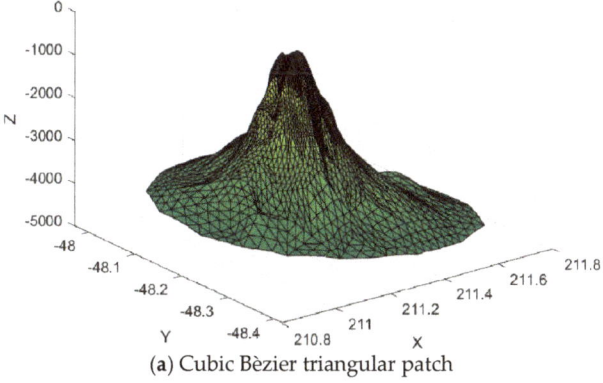

(**a**) Cubic Bèzier triangular patch

Figure 27. *Cont.*

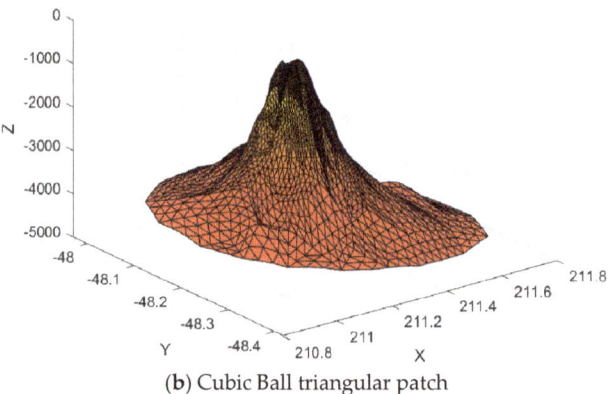

(**b**) Cubic Ball triangular patch

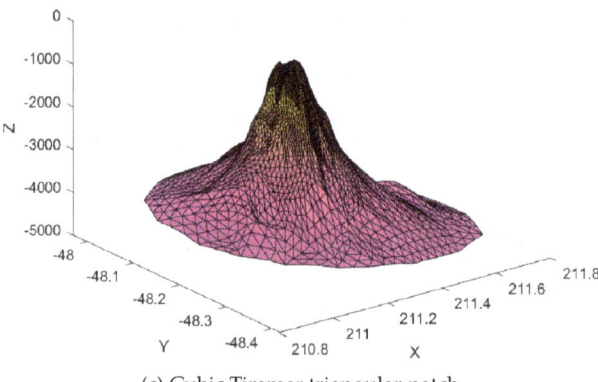

(**c**) Cubic Timmer triangular patch

Figure 27. Surface interpolation.

Then, we calculate the CPU time (in seconds) for each of the construction of the surfaces by using different methods shown in Figure 27. CPU time taken by proposed cubic Timmer triangular patches to construct the surface of the seamount dataset is 102.3931 s. Furthermore, cubic Bèzier and Ball triangular patches required about 103.7781 s and 103.5014 s, respectively. Based on this example, we conclude that the proposed cubic Timmer triangular patch required smaller CPU time especially for big or dense scattered datasets. Furthermore, based on Renka and Brown [22], when the coefficient of determination (R^2) is 0.999, the interpolation method can be considered as excellent. Therefore, from all numerical results, the proposed scheme is excellent.

6. Application on Real Data

In this section, the proposed scattered data interpolation using cubic Timmer triangular patch is tested to visualize some real datasets. We use two different datasets i.e., the rainfall data and the digital elevation data. All scattered data discussed in this section are irregularly distributed.

6.1. Visualize the Rainfall Data

First, we test the proposed scattered data interpolation scheme to visualize rainfall data. Based on our previous discussion, we apply the Foley and Opitz [19] method to calculate the inner ordinates and local scheme of Choice 1. The rainfall data sites are obtained from Malaysian Meteorology Department. The data are of average rainfall that were collected at some 25 major stations throughout Peninsular Malaysia. We have chosen the rainfall data for three different months i.e., February, March, and May 2007, as shown in Table 10. Figure 28 shows the Delaunay triangulation of rainfall data at the collected stations.

Table 10. Rainfall data.

Station	Location		Average Rainfall (mm)		
	Longitude	Latitude	Feb	March	May
Chuping	100.2667	6.4833	68.6	61.0	88.0
Langkawi Island	99.7333	6.3333	49.8	40.6	166.0
Alor Setar	100.4000	6.2000	99.4	277.8	67.4
Butterworth	100.3833	5.4667	31.6	58.9	143.2
Prai	100.4000	5.3500	33.8	208.1	153.4
Bayan Lepas	100.2667	5.3000	39.8	125.2	144.4
Ipoh	101.1000	4.5833	242.4	364.2	42.6
Cameron Highlands	101.3667	4.4667	117.2	252.0	223.2
Lubok Merbau	100.9000	4.8000	62.6	156.4	98.4
Sitiawan	100.7000	4.2167	49.8	44.4	26.8
Subang	101.5500	3.1167	199.0	329.2	68.2
Petaling Jaya	101.6500	3.1000	139.8	321.0	196.2
KLIA (Sepang)	101.7000	2.7167	78.6	186.2	188.8
Melaka	102.2500	2.2667	62.4	113.8	183.4
Batu Pahat	102.9833	1.8667	219.0	182.0	195.0
Kluang	103.3100	2.0167	39.4	92.4	130.2
Senai	103.6667	1.6333	176.3	148.6	296.0
Kota Bahru	102.2833	6.1667	7.0	115.2	109.2
Kuala Krai	102.2000	5.5333	16.6	166.0	238.7
Kuala Terengganu	103.1000	5.3833	0.2	121.0	64.8
Kuantan	103.2167	3.7833	35.2	79.2	270.4
Batu Embun	102.3500	3.9667	103.2	146.2	256.2
Temerloh	102.3833	3.4667	25.6	114.2	324.2
Muadzam Shah	103.0833	3.0500	85.0	131.6	204.8
Mersing	103.8333	2.4500	16.4	183.4	196.2

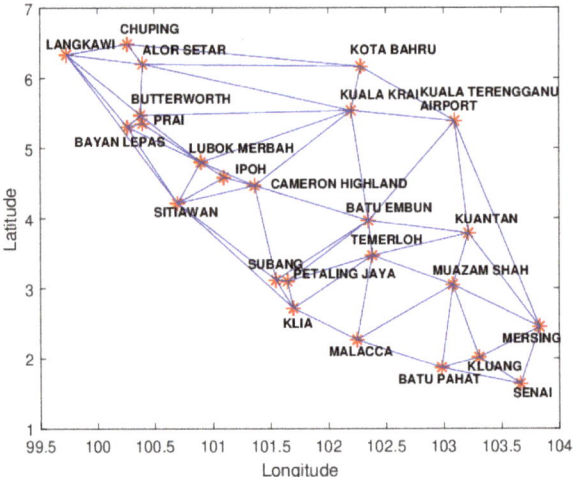

Figure 28. Delaunay triangulation of rainfall data.

Figure 29 shows the 3D linear interpolant of the rainfall data for each month. The surface of rainfall distribution in Malaysia of cubic Timmer triangular patches according to each month is shown in Figure 30.

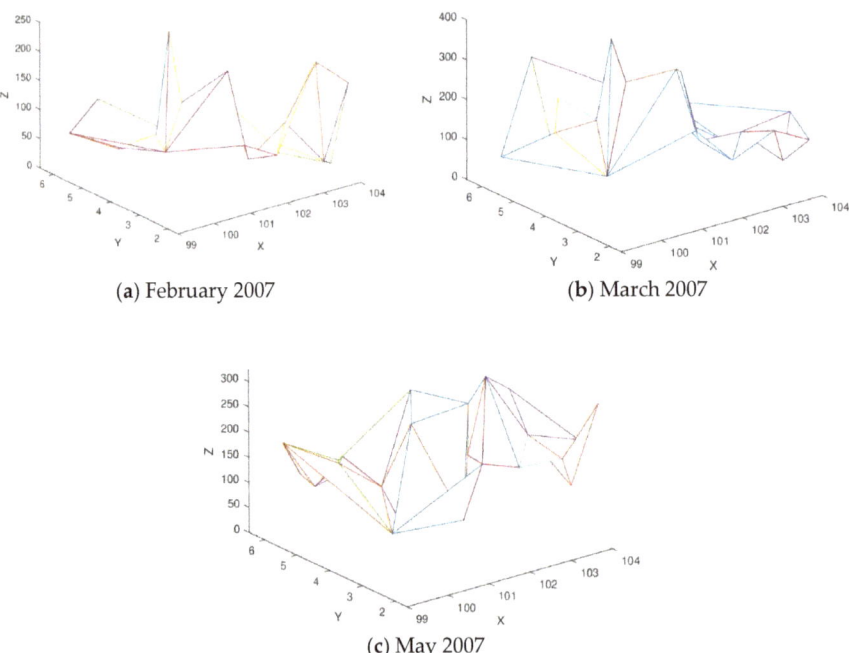

Figure 29. 3D Linear Interpolant of the rainfall data.

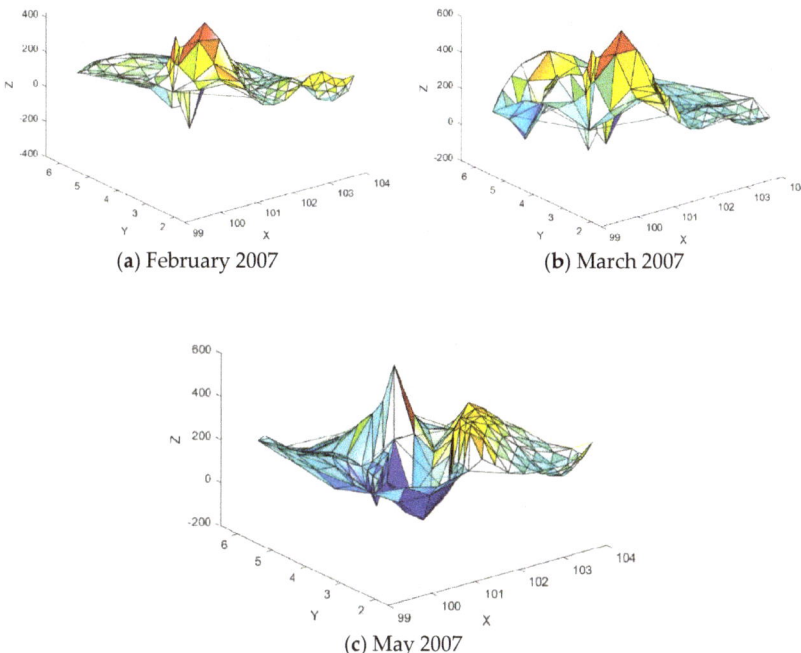

Figure 30. Surface Interpolation using the proposed scheme.

Meanwhile, Table 11 shows all the surface interpolations show that the minimum value of the average rainfall less than zero as shown in Table 11.

Table 11. Minimum value (mm).

Average Rainfall (mm)	Minimum Value (mm)
Feb	−259.9196
March	−165.0851
May	−71.3284

6.2. Visualize the Digital Elevation Data of Kalumpang Agricultural Station

Next, we test the proposed scattered data interpolation scheme to visualize the 160 digital elevation data of Kalumpang Agricultural Station (3° 38′ N, 101.34′ E) located about 90 km northeast of Kuala Lumpur, Malaysia (refer to z values in Table 12). We also apply the Foley and Opitz [19] method and Choice 1 to calculate the inner ordinates and local scheme of the proposed cubic Timmer triangular patch. Figure 31 shows the Delaunay triangulation for all 160 data points.

Table 12. Digital elevation data of Kalumpang Agricultural Station.

X	Y	Z	X	Y	Z	X	Y	Z
0	0	73.70	6	6	81.40	14	4	83.60
0	1	75.00	6	7	82.90	14	5	83.40
0	2	76.20	7	0	72.25	14	6	84.40
0	3	77.40	7	1	74.40	14	7	85.45
0	4	78.60	7	2	75.85	15	0	79.00
0	5	79.70	7	3	77.40	15	1	80.30
0	6	80.80	7	4	79.10	15	2	82.20
0	7	81.70	7	5	80.50	15	3	82.20
1	0	73.25	7	6	81.60	15	4	83.80
1	1	74.60	7	7	83.50	15	5	83.90
1	2	75.80	8	0	72.80	15	6	84.75
1	3	77.25	8	1	74.70	15	7	85.70
1	4	78.45	8	2	76.00	16	0	79.75
1	5	79.70	8	3	77.80	16	1	81.60
1	6	80.85	8	4	79.60	16	2	82.70
1	7	82.20	8	5	80.95	16	3	82.80
2	0	72.90	8	6	82.15	16	4	84.25
2	1	74.10	8	7	83.40	16	5	84.25
2	2	75.60	9	2	77.00	16	6	85.20
2	3	77.00	9	3	78.00	16	7	85.90
2	4	78.60	9	4	80.25	17	0	80.40
2	5	79.60	9	5	81.40	17	1	81.50
2	6	81.15	9	6	82.50	17	2	83.30
2	7	82.30	9	7	83.80	17	3	83.30
3	0	72.25	10	2	77.55	17	4	84.75
3	1	73.80	10	3	79.20	17	5	84.75
3	2	75.70	10	4	80.70	17	6	85.45
3	3	76.80	10	5	81.75	17	7	86.10
3	4	78.40	10	6	83.15	18	0	81.20
3	5	79.80	10	7	84.15	18	1	82.10
3	6	81.10	11	2	78.40	18	2	83.80
3	7	82.40	11	3	79.80	18	3	83.80
4	0	72.20	11	4	81.20	18	4	85.15
4	1	73.70	11	5	82.30	18	5	85.10
4	2	75.50	11	6	83.45	18	6	85.60
4	3	77.00	11	7	84.65	18	7	86.20
4	4	78.40	12	1	77.50	19	0	81.70
4	5	79.80	12	2	80.40	19	1	82.75
4	6	81.05	12	3	80.40	19	2	84.25
4	7	82.45	12	4	82.80	19	3	84.25
5	0	71.15	12	5	82.80	19	4	85.40
5	1	73.75	12	6	83.70	19	5	85.40
5	2	75.75	12	7	84.70	19	6	85.80
5	3	77.15	13	1	78.50	19	7	86.30
5	4	78.50	13	2	80.80	20	0	82.50
5	5	79.90	13	3	80.80	20	1	82.80
5	6	81.40	13	4	83.50	20	2	84.60
5	7	82.60	13	5	83.50	20	3	84.60
6	0	72.15	13	6	84.00	20	4	85.10
6	1	73.75	13	7	85.00	20	5	85.60
6	2	75.80	14	0	78.50	20	6	86.00
6	3	77.25	14	1	79.30	20	7	86.50
6	4	78.80	14	2	81.60			
6	5	80.20	14	3	81.10			

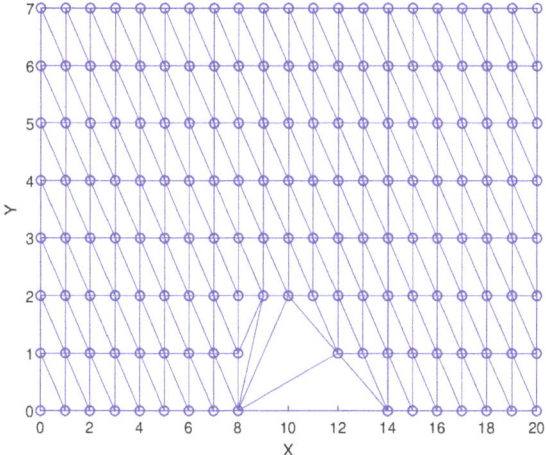

Figure 31. Delaunay triangulation for digital elevation data.

Figure 32 shows the 3D linear interpolant of the data points. The interpolating surfaces using the proposed cubic Timmer, Karim and Saaban [21], and Goodman and Said [8] schemes are illustrated in Figure 33. The final interpolating surface is constructed by combing all 269 triangular patches.

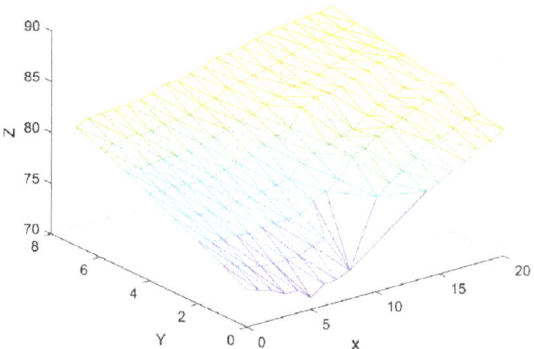

Figure 32. 3D linear interpolant for digital elevation data.

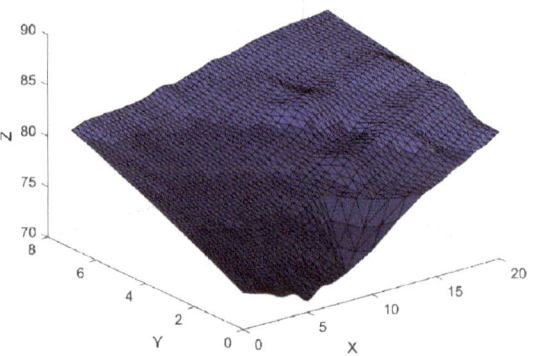
(a) The proposed cubic Timmer triangular scheme

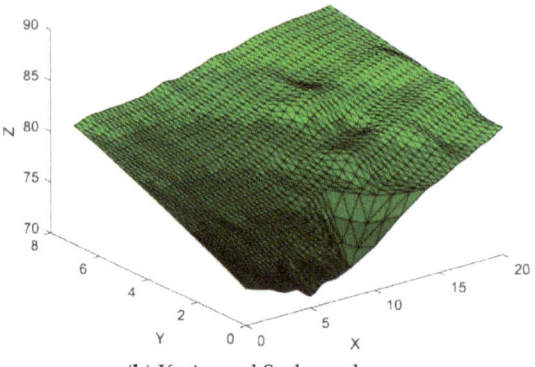
(b) Karim and Saaban scheme

(c) Goodman and Said scheme

Figure 33. Surface Interpolation.

Based on Figure 33, all interpolating surface are visually pleasing. However, we can compare the effectiveness of the proposed scheme by calculating the CPU time (in seconds). The proposed cubic Timmer triangular scheme took about 33.6956 s to construct the surface. Furthermore, the Karim and

Saaban [21] scheme required 33.8426 s while the Goodman and Said [8] scheme required 33.9239 s to interpolate the surface of digital elevation data. Hence, we can conclude that the proposed cubic Timmer triangular scheme is the best scheme in terms of CPU time compared to other schemes.

7. Conclusions and Future Work

In this paper, the cubic Timmer triangular patches, as implemented in Ali et al. [16], is applied to interpolate the scattered data. Goodman and Said [8] and Foley and Opitz [19] schemes are used in order to calculate the inner ordinates for each local scheme. It is observed that the cubic Timmer triangular patches offer lower CPU time (computational cost) as compared to the cubic Bezier and Ball triangular patches methods. Moreover, the simulation error shows that the cubic Timmer triangular patches have the same values as obtained from Goodman and Said schemes. In addition, we infer from the obtained results that the cubic Timmer triangular patches give better results as compared to some established schemes when the datasets are increasing. Therefore, we can preserve the positivity of the rainfall data by constructing the shape preservation of cubic Timmer triangular patches in our main studies in future.

Author Contributions: Conceptualization, S.A.A.K.; formal analysis, A.G. and D.B.; funding acquisition, S.A.A.K.; methodology, A.G. and K.S.N.; software, F.A.M.A., S.A.A.K., A.S., and M.K.H.; visualization, A.G.; writing—original draft, F.A.M.A., S.A.A.K., A.S., M.K.H., and K.S.N.; writing—review and editing, K.S.N. and D.B. All authors have read and agreed to the published version of the manuscript.

Funding: This study is fully supported by Universiti Teknologi PETRONAS (UTP) through its research grants YUTP:0153AA-H24 and the Ministry of Education, Malaysia through FRGS/ 1/2018/STG06/UTP/03/1015MA0-020.

Conflicts of Interest: The authors declare no conflict of interest.

References

1. Amato, F.; Moscato, V.; Picariello, A.; Sperlí, G. Recommendation in social media networks. In Proceedings of the 2017 IEEE Third International Conference on Multimedia Big Data (BigMM), Laguna Hills, CA, USA, 19–21 April 2017; pp. 213–216.
2. Karim, S.A.B.A.; Saaban, A. Visualization Terrain Data Using Cubic Ball Triangular Patches. In Proceedings of the MATEC Web of Conferences, 18–19 September 2018; Volume 225, p. 06023.
3. Ni, H.; Li, Z.; Song, H. Moving least square curve and surface fitting with interpolation conditions. In Proceedings of the 2010 International Conference on Computer Application and System Modeling (ICCASM 2010), Taiyuan, China, 22–24 October 2010; Volume 13, pp. V13–V300.
4. Ali, F.A.M.; Karim, S.A.A.; Dass, S.C.; Skala, V.; Saaban, A.; Hasan, M.K.; Ishak, H. Efficient Visualization of Scattered Energy Distribution Data by Using Cubic Timmer Triangular Patches. In *Energy Efficiency in Mobility Systems*; Sulaiman, S.A., Ed.; Springer: Singapore, 2020; pp. 145–180.
5. Awang, N.; Rahmat, R.W. Reconstruction of Smooth Surface by Using Cubic Bezier Triangular Patch in Gui. *Malays. J. Ind. Technol.* **2017**, *2*, 61–69.
6. Cavoretto, R.; Rossi, A.D.; Dell'Accio, F.; Tommaso, F.D. Fast computation of triangular Shepard interpolants. *J. Comput. Appl. Math.* **2019**, *354*, 457–470. [CrossRef]
7. Grise, G.; Meyer-Hermann, M. Surface reconstruction using Delaunay triangulation for applications in life sciences. *Comput. Phys. Commun.* **2011**, *182*, 967–977. [CrossRef]
8. Goodman, T.N.; Said, H. A Triangular Interpolant Suitable for Scattered Data Interpolation. *Commun. Appl. Numer. Methods* **1991**, *7*, 479–485. [CrossRef]
9. Hussain, M.Z.; Hussain, M. Shape preserving scattered data interpolation. *Eur. J. Sci. Res.* **2009**, *25*, 151–164.
10. Hussain, M.Z.; Sarfraz, M. Monotone piecewise rational cubic interpolation. *Int. J. Comput. Math.* **2009**, *86*, 423–430. [CrossRef]
11. Hussain, M.Z.; Hussain, M. C1 positivity preserving scattered data interpolation using rational Bernstein-Bézier triangular patch. *J. Appl. Math. Comput.* **2011**, *35*, 281–293. [CrossRef]
12. Karim, S.A.A. Monotonic Interpolating Curves by Using Rational Cubic Ball Interpolation. *Appl. Math. Sci.* **2014**, *8*, 7259–7276. [CrossRef]

13. Karim, S.A.A.; Saaban, A.; Hasan, M.K.; Sulaiman, J.; Hashim, I. Interpolation using Cubic Bèzier Triangular Patches. *Int. J. Adv. Sci. Eng. Inf. Technol.* **2018**, *8*, 1746–1752. [CrossRef]
14. Ibraheem, F.; Hussain, M.Z.; Bhatti, A.A. C^1 Positive Surface over Positive Scattered Data Sites. *PLOS ONE* **2015**, *10*, e0120658. [CrossRef] [PubMed]
15. Su, X.; Sperli, G.; Moscato, V.; Picariello, A.; Esposito, C.; Choi, C. An Edge Intelligence Empowered Recommender System Enabling Cultural Heritage Applications. *IEEE Trans. Ind. Inform.* **2019**, *15*, 4266–4275. [CrossRef]
16. Ali, F.A.M.; Karim, S.A.A.; Dass, S.C.; Skala, V.; Saaban, A.; Hasan, M.K.; Ishak, H. New cubic Timmer triangular patches with C1 and G1continuity. *J. Teknol.* **2019**, *81*, 1–11.
17. Timmer, H.G. Alternative representation for parametric cubic curves and surfaces. *Comput.-Aided Des.* **1980**, *12*, 25–28. [CrossRef]
18. Goodman, T.N.T.; Said, H.B.; Chang, L.H.T. Local derivative estimation for scattered data interpolation. *Appl. Math. Comput.* **1995**, *68*, 41–50. [CrossRef]
19. Foley, T.A.; Opitz, K. Hybrid cubic Bézier triangle patches. In *Mathematical Methods in Computer Aided Geometric Design II*; Academic Press: New York, NY, USA, 1992; pp. 275–286.
20. Awang, N.; Rahmat, R.W.; Sulaiman, P.S.; Jaafar, A. Delaunay Triangulation of a missing points. *J. Adv. Sci. Eng.* **2017**, *7*, 58–69.
21. Karim, S.A.A. Shape Preserving by Using Rational Cubic Ball Interpolant. *Far East J. Math. Sci.* **2015**, *96*, 211–230.
22. Renka, R.J.; Brown, R. Algorithm 792: Accuracy Tests of ACM Algorithms for Interpolation of Scattered Data in the Plane. *ACM Trans. Math. Softw.* **1999**, *25*, 78–94. [CrossRef]

© 2020 by the authors. Licensee MDPI, Basel, Switzerland. This article is an open access article distributed under the terms and conditions of the Creative Commons Attribution (CC BY) license (http://creativecommons.org/licenses/by/4.0/).

MDPI
St. Alban-Anlage 66
4052 Basel
Switzerland
Tel. +41 61 683 77 34
Fax +41 61 302 89 18
www.mdpi.com

Mathematics Editorial Office
E-mail: mathematics@mdpi.com
www.mdpi.com/journal/mathematics

www.ingramcontent.com/pod-product-compliance
Lightning Source LLC
LaVergne TN
LVHW071954080526
838202LV00064B/6745